SELECTED PAPERS OF JOHN H. HOLLAND

A Pioneer in Complexity Science

Exploring Complexity

For four centuries our sciences have progressed by looking at its objects of study in a reductionist manner. In contrast complexity science, that has been evolving during the last 30–40 years, seeks to look at its objects of study from the bottom up, seeing them as systems of interacting elements that form, change, and evolve over time. Complexity therefore is not so much a subject of research as a way of looking at systems. It is inherently interdisciplinary, meaning that it gets its problems from the real non-disciplinary world and its energy and ideas from all fields of science, at the same time affecting each of these fields.

The purpose of this series on complexity science is to provide insights in the development of the science and its applications, the contexts within which it evolved and evolves, the main players in the field and the influence it has on other sciences.

Volume 4: *Selected Papers of John H. Holland: A Pioneer in Complexity Science*
edited by Jan Wouter Vasbinder & Helena Hong Gao

Volume 3: *43 Visions for Complexity*
edited by Stefan Thurner

Volume 2: *Cultural Patterns and Neurocognitive Circuits: East–West Connections*
edited by Jan Wouter Vasbinder & Balázs Gulyás

Volume 1: *Aha.... That is Interesting! John Holland, 85 Years Young*
edited by Jan Wouter Vasbinder

Exploring Complexity – Volume 4

SELECTED PAPERS OF JOHN H. HOLLAND

A Pioneer in Complexity Science

Editors

Jan Wouter Vasbinder
Helena Hong Gao
NTU, Singapore

World Scientific

NEW JERSEY • LONDON • SINGAPORE • BEIJING • SHANGHAI • HONG KONG • TAIPEI • CHENNAI • TOKYO

Published by

World Scientific Publishing Co. Pte. Ltd.
5 Toh Tuck Link, Singapore 596224
USA office: 27 Warren Street, Suite 401-402, Hackensack, NJ 07601
UK office: 57 Shelton Street, Covent Garden, London WC2H 9HE

British Library Cataloguing-in-Publication Data
A catalogue record for this book is available from the British Library.

Exploring Complexity — Vol. 4
SELECTED PAPERS OF JOHN H. HOLLAND
A Pioneer in Complexity Science

ISBN 978-981-3234-76-5

For any available supplementary material, please visit
http://www.worldscientific.com/worldscibooks/10.1142/10841#t=suppl

Printed in Singapore

Preface

John Holland died on August 9th 2015. He was a very imaginative and creative thinker who once remarked: "I have more ideas than I can ever follow up on in a lifetime, so I never worry about someone stealing an idea from me." While many of the John's ideas are greatly influencing many fields of science, many more of his ideas have never been followed up.

At a meeting with John Holland, Professor Kok Khoo Phua, the Chairman of Word Scientific (WS) in Singapore and the founding Director of the Institute of Advanced Studies at Nanyang Technological University (NTU), convinced John that the publication of a selection of his papers might have a great influence on research in the world and especially in Singapore.

John had an intense relationship with Singapore, and specifically with NTU. Since 2007 he was a frequent visitor to NTU, initially to do joint research with Helena Gao on language acquisition from the perspective of complex adaptive systems, and later also to attend (and be a speaker at) the conferences organized by the complexity programme directed by Jan W. Vasbinder.

In 2009 NTU organized a conference in honor of John Holland's 80th birthday and to emphasize his close ties with NTU.

During his visits to NTU over the years, John delivered a course on complexity adaptive systems to the graduate students, organized many different workshops with senior scientists, young researchers and graduate students. Each time he inspired the participants with new ideas, or with new links between theories and current issues covering many fields. The discussions during these workshops made us realize that the theories that John developed in the 1960s and 1970s are highly applicable to current research issues.

Because of these discussions, young people started to look for his publications to see how powerful these theories are. To facilitate this, we started to look into his publications and found that his earliest publications are not easily available. At the same time, these publications entail a lot of explored and unexplored ideas. So, we decided that it would be of value to young (and old) scientists to make these publications available and provide young people a view into the enormous imagination and creativity of John Holland.

This collection of John Holland's papers represents his work in the field of computer logic from late 1950s to 1970s. It is not too much to say that with his work, and especially the work he did together with Arthur Burks, John laid the foundation under what is now called computer science.

Jan Wouter Vasbinder
Editor in Chief

Helena Gao
Co-editor in Chief

Acknowledgments

It was no easy task to select 15 papers on computer logic from among 13 boxfulls covering a great variety of subjects. The work could only be accomplished through the efforts of many people, who granted funding or access to the papers, or who organized the logistics of the process, or who examined and photographed the papers page by page.

We here wish to express our gratitude to: Prof. Bertil Andersson, President of Nanyang Technological University (NTU) in Singapore, who has always supported John H. Holland's interactions with NTU and who, from the very start, supported the idea of compiling a series of Holland's selected papers; Para Limes@NTU, that provided the travel support for the Holland paper selection project; Prof. Kok Khoo Phua, who took the initiative and then convinced John that it was a good idea to publish a series of selected papers; Prof. Alan Chan, for his support and recommendation of Mr. Jonathan Sim, his research assistant,to take the project trip with Dr. Helena Gao to the Bentley Historical Library at the University of Michigan for the search of the papers, the Bentley Historical Library at the University of Michigan that provided access to and permission to publish John H. Holland's academic archives from 1950s to 1970s, and Ms. Aprille McKay and Ms. Malgosia Myc in particular for their assistance and guidelines in the search of the catalogues; John's two daughters Gretchen and Alison and their families who supported the project by sharing their memories of their father's research interests; Mr. Jonathan Sim who photographed all the papers; Mrs. Veronica D. Lampman who offered her great help on the last day of the paper search in the library; Ms. Lakshmi Narayanan, who managed the project from World Scientific publishers and showed almost endless patience, and finally Ms. Karen Chung, who provided the necessary administrative and logistics back-up for the project from Para Limes.

Contents

2900-52-R

Memorandum of Project MICHIGAN

SURVEY OF AUTOMATA THEORY

John Holland

September 1959

The University of Michigan
WILLOW RUN LABORATORIES
Ann Arbor, Michigan

CONTENTS

ABSTRACT

Automata theory (presently more formally named, at M. I. T. and The University of Michigan, "The Communication Sciences") draws its methods and data from several disciplines. Major contributions of four disciplines, and the way in which these contributions are integrated to form a unified discipline, are surveyed here. The four disciplines are: (1) logic and metamathematics, (2) the theory and use of stored-program computing machines, (3) information theory, and (4) neurophysiology and related parts of psychology. From logic and metamathematics come the concepts of a formal system (the axiomatic method and decision procedures), general recursive functions (effective definition), and Turing machines (computability). Study of stored-program computers contributes the concepts of the logical net, programming, and simulation. Information theory provides a means of studying complex systems by statistical methods, giving a measure of the amount of data a system can process or store, and rigorous definitions to concepts such as redundancy and feedback. Finally, neurophysiology and psychology supply data and examples of very complex systems, important questions, and some suggestive outlines of theory. Consideration is given to the way in which these methods and concepts are coordinated to provide abstract deductive theories of complex systems with delayed feedback.

1
INTRODUCTION

At the present time the systems used in communication, data processing, and control are undergoing a tremendous increase in complexity and capability. A new discipline, whose object is to explore the nature and range of action possible for these systems, has been developing. This discipline — referred to here as automata theory — operates at the level of theory and yields information about these new systems in much the same way that theoretical physics yields information about what is possible within the realm of physics. One of the objectives of the discipline is a theory of systems having a variable structure; thus it is pertinent to research in several separate fields, including error-correcting or self-repairing systems, automatic programming, artificial intelligence, and genetics.

The present memorandum singles out major contributions in four areas of study and surveys the way in which these contributions are integrated to form the new discipline. Although the selection of topics is meant to be representative, the survey is not intended to be comprehensive in its selections, either of the contributing areas of study or of the contributions of these areas. The survey _is_ intended to provide those familiar with some part of one or more of the areas mentioned with a general background which will help them to understand the way in which automata theory provides information about the potentialities of systems of the sort mentioned above.

Section 2 of the survey discusses some of the general characteristics of the new discipline. Section 3 introduces four disciplines bordering automata theory and gives a suggestion of the way in which these four disciplines interact in automata theory. These disciplines are: (1) logic and metamathematics, (2) the theory and use of stored-program computing machines, (3) information theory, and (4) neurophysiology and related parts of psychology. Sections 4 through 7 discuss each of these disciplines in turn. Section 8 concludes the survey by outlining a possible approach, using the techniques of automata theory, to a very general problem of interest to several areas: the problem of self-organizing systems.

2
COMMENTS ON UNITY

When a new area of study emerges from the range of established disciplines, questions of unity and integration follow directly. Such questions are natural and important, being prompted by interests akin to those which start the hunt for laws to relate a body of similar facts. In the present case, unity is a special concern because, in each of several established disciplines, one can find particular instances of problems, concepts, or methods characteristic of the new area. For this reason, effort must be exerted toward showing that the area is not just an _ad hoc_ collection of problems, methods, and so on.

The discussion is encumbered at present by the lack of an accepted name for the area; the names "cybernetics," "information theory," "science of computing machines," "automata theory," and "communication sciences" are currently applied, in overlapping fashion, to major portions of the area. In order to go more directly to the questions of unity and integration, this problem will be disposed of for now by referring to the area as automata theory.

From the outset it is apparent that concepts of complex natural and artificial systems underlie the whole area. But it is the way in which these concepts are formulated and developed that really characterizes the area. Abstract deductive theory plays an indispensable part, since automata theory aims primarily at developing deductive theories of the complex systems involved. A major part of any investigation is carried on at the theoretical level, and many investigations take place wholly at the theoretical level. In the latter cases, the abstract theory characterizes or idealizes some very broad class of complex systems about which general knowledge is desired. In all cases, the theory is based (either explicitly or implicitly) upon the concept of the system being, at each moment, in one of a well-defined set of states. Usually this set is finite at each moment and quite often time is quantized. The resulting abstraction is called a discrete system. (Even with continuous signals, as in parts of information theory, when sampling theorems and a fidelity criterion are used, one has, in effect, a discrete system.) In general, the state of the system (sometimes called the complete state) can be decomposed into two disjoint parts — the internal state and the input state. The input state is determined by conditions independent of the system. (In some cases, the set of input states may be empty.) The succession of internal states is then determined by a finite (or effective) set of rules. The rules specify the internal state at any given time in terms of the input and internal states over a related (usually earlier) set of times. Often, in order to treat systems by the methods of statistics (a major contribution of information theory), at each moment a set of "possible" input states is distinguished and each of these states is assigned a probability of occurrence. In this case, the rules specify the probabilities of the internal states. Rules, of either the former or latter type, together with various postulates determining the number of states, provide the basis for any deductive theory of complex systems. The resulting abstract deductive theories pervade automata theory; they are a vital integrating factor.

A few examples of such theories, as applied to distinctly different complex systems, will show just how general their use is within automata theory. Chomsky (Ref. 21) makes a typical application of this approach to the study of language. He defines a finite state language as a set of strings (sentences) of symbols (words) generated by a finite set of rules. The system can be in one of a finite number of states at any given moment. Each rule specifies the state of the system in which it can be applied, the symbol which is generated, and the new state of the generator after the rule has been applied. A second example results from application of the technique to natural systems with intricate patterns of delayed feedback. This usually begins with an analysis of the system into units. The functional characteristics of these units are fully defined so that for each possible input to a unit, the output is completely specified. The investigation proceeds by examining the behavior (succession of states) in various systems constructed by connecting these units. The work of Rochester et al. (Ref. 10) is an example of this method as applied to the study of the central nervous system. A similar procedure is used to study digital computing circuits. A final, very general, example is the Turing machine (Ref. 6). In its original form, this machine consists of a control unit which scans a sequence of storage locations (a "tape"). The control unit can be in any one of a finite number of states, and various symbols can be stored at each of the locations (the "program"). There is a set of rules which causes the control unit, for each scanned-symbol/control-state combination, to store a given symbol at the scanned location, change state, and scan a new location. To start the machine, the control unit is set to scan one of the storage locations and given an initial state. In a certain sense, a Turing machine can approximate to arbitrary precision the action of any complex natural or artificial system. Each of these systems will be considered in greater detail later, but for the present these brief descriptions will perhaps serve. In each case, the deductive theory is built around the concept of a set of states and a finite set of rules which specify the possible sequences of these states. Because of the latitude of description possible within such a framework, automata theory is characterized much more by its methods than its subject matter.

Although development in automata theory is primarily theoretical, physical systems are always the ultimate source of definitions, axioms, rules, etc. Physical systems also serve as final arbiter of the value of any development insofar as it is aimed at results in this area. That is, so far as a given theory increases understanding of some class of complex systems, so far is it valuable in this area; and so far as the former is doubtful, just so far is the latter. Generally, the succession of states as determined by the rules in the deductive theory will correspond in a significant way to the succession of conditions in the physical system. The abstract theory will predict, at least approximately, the action of the physical system. If the hypotheses and rules of such a theory are general enough, then any theorems derived will hold for large classes of complex systems. For example, consider a theory which deals with a set of basic units which are to be connected in various ways to represent different physical systems. Then statements about structure will logically imply statements about behavior. Such deductions about behavior will hold true for any concrete system for which the structural statements hold. In all this, automata theory plays much the same role with respect to complex systems that theoretical physics plays with respect to the material systems of physics.

3
BORDERING DISCIPLINES

Three disciplines have contributed substantially to the methods of automata theory: (1) logic and the foundations of mathematics, (2) the theory and use of stored-program computing machines, and (3) information theory. From logic and foundations come the concepts of a formal system (the axiomatic method and decision procedures), general recursive functions (effective definition), and Turing machines (computability). From the study of computing machines come the ideas of the logical net, programming, and simulation. From information theory comes a statistical measure of the amount of data a system can process or store, and rigorous definitions of such concepts as redundancy and feedback.

These methods are combined in various characteristic ways to provide automata theory with tools for the theoretical investigation of complex systems. In a typical combination, a formal system is built around a set of recursive equations which give the functional characteristics and the connection scheme of a set of basic units abstracted from some complex natural system. This set of equations is, in effect, a set of rules which determines the succession of states (or some properties thereof) in the abstract system. Special cases of the formal system can then be programmed for simulation on a computer. The program can be run with different conditions placed upon the recursive functions corresponding to changes in feedback, redundancy, and so on, in the natural system. The data thus accumulated will often suggest general theorems about feedback or redundancy in the natural system. Then, returning to the formal system, one can attempt to prove the suggested theorems in a rigorous way. A more extended discussion of the interplay and integration of methods from the three fields will be given as the fields themselves are discussed in the next three sections.

One other discipline will be included in this survey: neurophysiology and the related parts of psychology. By contributing data and suggestions for a theory of the central nervous system — one of the most versatile and complex natural systems known to man — this descipline has provided several well-defined problems and an important subject matter for automata theory. The artificial intelligence problem, a problem of tremendous interest both theoretically and practically, was suggested by this field (in conjunction with the advent of stored-program computers). Its questions are ones like: How does one construct a system which will learn, form concepts, or model its environment? The fact that this problem requires methods and data from several established disciplines prevents an investigation within the boundaries of any one of them. However, it does lie squarely within the domain of automata theory. Because of its importance, and because its exploration requires the whole range of methods available to automata theory, this problem by itself comes close to justifying the emergence of the new discipline.

4

LOGIC AND FOUNDATIONS

4.1 FORMAL SYSTEMS AND LOGIC CALCULI

The concept of a formal system, which in some aspects has been with us since the Fourth Century B. C., has in the past century undergone far-reaching refinements and revisions. According to the present view, a formal system consists of four parts: a set of primitives, a definition of what constitutes a term or well-formed formula, a set of rules, and a set of axioms. Each part must be effectively defined (see Sec. 4.2); for example, it must be possible to decide in a finite number of steps whether a given term is an axiom. Repeated application of the rules to the axioms will generate a set of terms which are called consequences of the axioms. One can give an inductive definition (see Sec. 4.2) which will specify when the generation (deduction) of a given term is admissible.

Having set up a formal system, one can proceed to examine it at some length without paying any direct attention to its content. An important tool in this investigation is the valuation. A valuation is a mapping from the terms of the formal system to some domain of individuals (and to Cartesian products thereof). A valuation is said to satisfy the formal system when the mapping satisfies criteria, inductively given, which correspond to the intuitive notion of a model in which the admissible terms are true. (In automata theory, one always has some set of complex systems in mind, and the axioms are chosen so that upon interpretation they become true statements about important or characteristic properties of the systems.) A term is m-satisfiable if there is a domain of cardinality m and a valuation on this domain such that the term is satisfied. If a term is satisfied by all valuations in a domain of cardinality m, it is m-valid; the term is valid if it is m-valid for all m. With the aid of these concepts, one can investigate the completeness (all valid terms are admissible), consistency (some term is not admissible), and axiom independence of the formal system. The question of consistency is crucial and in many cases is closely bound up with the question of completeness (see Gödel's theorem). In some cases, the most that has been proved for a system is its relative consistency (there is a mapping such that the given system is satisfied by a fragment of a second formal system). Thus certain non-Euclidean geometries have been proved consistent relative to Euclidean geometry.

One of the most important examples of a formal system comes from logic. It is the first-order function calculus. The primitives are symbols for individual variables, predicate symbols, and a set of logical connectives, together with a quantification symbol and parentheses. Deciding when a set of primitives is sufficient to allow the generation of a complete first-order function calculus is a question of some interest. (See the analogous problems of choosing the set of operations for a Turing machine or primitive elements for logical nets.) The well-formed formulas are inductively defined with such strings as $\sim Vx_1x_1$, etc., ruled out. Various combinations of axioms and rules are possible — the idea being to achieve completeness. It is worth noting that for the propositional calculus, a fragment of the first-order calculus with no predicates or quantifiers, deductive completeness serves (either a term or else its negation is deducible). For the first-order function calculus, however, the stronger notion of completeness in terms of validity is required.

Once the logic calculus has been set up, metatheorems relating to the consistency and completeness of the system receive priority. For the propositional calculus and the first-order functional calculus, these tasks were considerably simplified by the discovery of certain normal forms for the terms: the disjunctive and Boolean normal forms in the case of the propositional calculus; the prenex and Skolem normal forms in the case of the first-order function calculus. In addition, for the propositional calculus, there is the truth-table method (closely tied up with the Boolean normal form) of deciding whether any given term is a tautology (admissible). Procedures of this kind (decision procedures) turn out to be most important to metamathematics and automata theory.

4.2 RECURSIVE FUNCTIONS

Recursive functions are a formalization of the general mathematical procedure of definition by induction. (Recursive functions and definitions occur in all parts of automata theory. The use of recursive definitions in formal systems to decide when a formula is well formed or a deduction valid has been noted. Recursive functions play an analogous role in the theory of logical nets, and, in fact, one can use a set of recursive equations to represent a logical net. In programming, recursion or iteration is almost the <u>sine qua non.</u>) The simplest class of recursive functions are the primitive recursive functions. A primitive recursive function is defined in terms of a set of initial functions and applications of the schema of substitution and recursion to these initial functions. The initial functions can be of three kinds: (1) $\phi(x) = x'$ (intuitively $\phi(x) = x + 1$), the successor function, (2) $\Phi(x_1, \ldots, x_n) = q$, the constant function, and (3) $\Phi(x_1, \ldots, x_i, \ldots, x_n) = x_i$, the identity function. The schema of substitution can be symbolized as

$$\phi(x_1, \ldots, x_n) = \psi[\chi_1(x_1, \ldots, x_n), \ldots, \chi_m(x_1, \ldots, x_n)] .$$

The schema of recursion can be symbolized as

$$\phi(0, x_2, \ldots, x_n) = \psi(x_2, \ldots, x_n)$$

$$\phi(x', x_2, \ldots, x_n) = \chi[x, \ (x, x_2, \ldots, x_n), x_2, \ldots, x_n] .$$

Having defined primitive recursive functions, one can extend the idea of primitive recursion to other objects; e.g., a primitive recursive predicate is one which has a primitive recursive representing function.

One of the most important characteristics of the primitive recursive functions is that for every value of the argument there is an explicit procedure for constructing the value of the function. This leads directly to the idea of an effective function; $y = f(x)$ is an effective function if for every x for which the function is defined there is an explicit procedure for producing y. Similarly, a predicate, P(x), is effective if there is procedure whereby one can decide for every x whether P(x) is true, i.e., a decision procedure. (In automata theory all systems are in principle constructible and the state or probability distribution of states is effectively defined as a function of time and the input states or distributions.) If an effective predicate is quantified, then the result, in general, is a semieffective predicate. An important example of these two cases is the predicate $D(F, y, x) \equiv$ "the sequence y is a proof of x in the formal system F," where y is any finite ordered sequence of terms in F. The predicate D is effective. However, the predicate $T(F, x) \equiv (\exists y)D(F, y, x) \equiv$ "x is a theorem in F" is in general semieffective. That is, one may find a proof for x, thus showing that x is a theorem; but x may not be a theorem, in which case the search for a proof is hopeless. Closely related to the idea of a semieffective predicate is an effectively enumerable predicate. A predicate, P(x), is effectively enumerable if there is an effective function, f(y), such that for each x which makes the predicate true there is an argument y for which $f(y) = x$. Note that P(x) is effective only if both P(x) and $\sim P(x)$ are effectively enumerable. From all this, one can see how the law of the excluded middle fails when interpreted in an effective manner. It is just not the case that for expressions such as $(\exists x)P(x)$ <u>either</u> there is an effective procedure for constructing an x such that the expression holds <u>or</u> there is an effective proof that $(\exists x)P(x)$ is false.

The primitive recursive functions are countable and can be ordered. Thus the possibility presents itself of forming new effective functions by using the diagonal procedure on an ordering of the primitive recursive functions. It turns out that in this way effective functions are generated which are not primitive recursive. However, the class of primitive recursive functions can be enlarged to include these new effective functions. The result is the class of general recursive functions. The class of general recursive functions has in fact been defined by means of a formal system, of which the central features are the rules of inference. The first rule permits one to substitute a numeral for a variable. The second rule in effect permits the substitution of equals for equals. A general recursive function is then given by a set of equations and its values are deduced by repeated application of the rules of inference. Kleene (Ref. 5) has shown that every general recursive function ϕ can be put in the normal form $\Phi = \psi[(\epsilon y)R]$, where ψ and R are primitive recursive and ϵ is the smallest number operator. It is interesting to note that, if the diagonal procedure is applied to a listing of general recursive functions, the resulting functions are not effectively defined.

One of the high points in the development of the theory of general recursive functions was Gödel's proof (Ref. 1,5) of the undecidability of certain terms in any system containing elementary arithmetic. Elementary arithmetic can be formalized by modifying ("making an application of") the first-order function calculus. First, only the predicates +, $\cdot$, =, ' (successor), and 0 (a zeroth-order predicate, i.e., a constant) are allowed as primitives. Then axioms are added to the calculus specifying properties of these predicates (commutivity, distributivity, etc.) corresponding to those usually used to define plus, times, equals, etc. In the resulting formal system Z one can then define numeralwise expressible predicates: the number predicate $P(u_1, \ldots, u_n)$ is numeralwise expressed by the formula $\alpha(x_1, \ldots, x_n)$ in Z if, for all numbers $p_1, \ldots, p_n$, the truth of P implies α is deducible in Z and the falsehood implies $\sim\alpha$ deducible. It can then be proved that a predicate is numeralwise expressible if and only if it is general recursive. Thus Z contains the theory of general recursive functions. By the process of Gödel-numbering, one can make a 1 to 1 correspondence between equations, predicates (such as the predicate D used above), etc, and specific numbers in Z. Now, by making an adaptation of the liar's paradox, Gödel was able to exhibit a term which is deducible if and only if its negation is deducible. This term, chosen with the help of a diagonal argument, has Gödel number h and corresponds to the metamathematical statement "the formula with Gödel number h is not deducible." Amazingly enough, the term asserts its own undecidability; thus, if Z is consistent, it is incomplete. (Variations of Gödel's proof can be used to show that in certain formal systems used to describe complex systems the problem of finding a general synthesis algorithm is unsolvable.)

4.3 TURING MACHINES

The Turing machine originated in A. M. Turing's investigation of computable numbers and computability (Ref. 6). Intuitively, computable numbers are expressions calculable by finite means. In order to give a rigorous formulation to this idea, Turing conceived of the calculation as being carried out by a machine. A tape, divided into squares and indefinitely extendible, supplies instructions to the machine and records the results of the computation. Symbols of two kinds are recorded on the tape: symbols representing numerical digits and symbols in effect representing orders to the machine. The machine itself is considered to have only a finite number of conditions or states. At any moment, it scans one square of the tape, and under certain conditions it may cause a symbol to be printed on the tape. These conditions and the actions of the machine as a whole are given by means of a table — the machine description. The first column gives the state of the machine. The second column gives the symbol on the tape square being scanned (which may, of course, be blank). The third column specifies what action the machine is to take as the result of the configuration of the first two columns. The action consists of printing a symbol (or not printing) and moving the machine to one of the adjacent squares on the tape. The fourth column gives the new state of the machine (the state the machine will be in when it scans the next square). Now any computable sequence will be considered defined by the machine description of the machine which will compute it.

The machine description of any machine can be reduced to a standard description employing, repeatedly, only seven symbols, e.g., the state q_i, is represented by DA ... A, where A is repeated i times. If the symbols are numerals, the result is a number called the description number of the machine. With the help of this result, Turing shows that there exists a machine, called the universal Turing machine, which when supplied with the description number and tape for any given machine will compute the same sequence as that machine. (This result indicates the possibility of using a stored-program computer to simulate other complex systems. This will be discussed in the section on stored-program computers.) Then, by a diagonal argument analogous to the one Gödel used, Turing proves that the computable numbers are enumerable and that the problem of deciding whether a given machine will stop is undecidable.

Questions of how many states, or symbols, or operations one needs to specify a universal Turing machine arise naturally at this point. Shannon has shown that a universal machine with only two states can be constructed. It is also possible to construct a universal machine with only two symbols (a "blank" and a "mark"). As yet, there is no result which indicates what the minimum (state) x (symbol) product is. However, the results quoted give weight to the concept of the interchangeability of symbols and states ("speed" and "storage"; see Sec. 5).

5
STORED-PROGRAM COMPUTING MACHINES

5.1 THEORY OF LOGICAL NETS

At present, there are two distinct approaches to the theory of stored-program computers: the theory of Turing machines (which was discussed in Sec. 4.3) and the theory of logical nets. The theory of Turing machines seems more closely related to programming and simulation (to be discussed presently), whereas the theory of logical nets is more concerned with structure and organization. The theory of logical nets (Ref. 9, 19) is based upon the idea of a set of primitive elements ("components") which can be connected in various ways. In one version of the theory, there are just two types of elements, the stroke element and the delay element. Each type of element is thought of as having a fixed number of inputs and one output. The stroke element has two inputs, the delay element one. At each moment of time, each input and each output has one of a fixed number of states — in the present case two, 0 or 1. The moments of time are thought of as countable and successive and are thus put into a 1: 1 correspondence with the natural numbers 0, 1, . . . , t, ... The output state of the stroke element at any moment, t, is given by the Sheffer stroke function (or the corresponding truth table) which has as arguments the states of the two element inputs at time t. The output state may be thought of as given by the predicate $S_i(t) \equiv I_j(t) \,|\, I_k(t)$, where $I_j(t)$ and $I_k(t)$ are predicates corresponding to the two inputs to the stroke element. In a similar way, the output of a delay element is given by $D_i(t + 1) \equiv I_j(t)$, where I_j is the predicate corresponding to the delay input.

Once the primitive elements have been specified, the next step is to lay down the rules of connection. This is done by giving a recursive definition of the class of well-formed nets. The result is that, for each computing device constructed of a fixed number of switches and storage elements, there is a well-formed net which mirrors it logically (and, to an extent, structurally). Actually, there are many alternative sets of primitives which, with appropriate connection rules, will yield essentially the same class of logical nets. Logical designers commonly use a set consisting of "and," "or," "not," and "delay" elements; McCulloch and Pitts, and Kleene (Ref. 22) have used a set of threshold elements with built-in delay; Burks and Wang (Ref. 9) have used an infinite class consisting of elements corresponding to each truth function together with a delay element. In each case, the process of connecting components is mirrored logically by identifying the output predicate of one element with the input predicates of the elements to which it is connected. If every input predicate is identified with some output predicate, the net is said to be input-free. Otherwise, the input predicates left unidentified with output predicates are designated net inputs. The values of the net input predicates cannot be determined from the net equations, but must be given independently. The resulting set of equations, the element equations and the net input equations, completely specifies the behavior of the logical net. The equations can be thought of as a set of recursions on the variable t.

If the output predicates of the delay elements in a net are ordered, then one can order their output states at any time t in the same way. The resulting sequence of 0's and 1's can be thought of as a binary number. This number at each time t is called the (coded) internal state of the net at time t. In a similar way, one can order the net input predicates and define a net input state. Now it can be shown that the internal state at time t + 1 is completely determined by the input and internal states at time t. A graph which represents the behavior of the logical net, the state transition diagram, can be drawn using this information. The vertexes of the graph represent internal states, and the edges, which go from each internal state to its successor for each net input state, represent net input states. Various interesting relations between structural properties of logical nets and the state transition diagrams can be exhibited. For example, one can define for logical nets the concept of a cycle (feedback loop) in the structure. Now when the input sequence to a net is periodic, the internal state sequence must also be. The most important influence on the relation of the internal to the input period (for nets having the same number of delay elements) is the number of cycles in the net. Another important property of logical nets can be deduced by the fact that the output sequence of an element is periodic if that sequence is input-independent, i.e., the same for all net input sequences. Thus no logical net can realize the primitive recursive function $0(t) \equiv (t = n^2)$, for this predicate is input-independent but aperiodic. This indicates that logical nets are weaker than Turing machines, which realize all recursive (computable) functions. However, the concept of a growing logical net, in which the number of elements is not fixed, provides a class of nets broad enough to include Turing machines.

5.2 COMPUTERS AND PROGRAMMING

Existing stored-program computers are made up of four major types of units: input-output units, operations units, storage units, and control units. The input-output units (punched card readers and punches, magnetic tape devices, printers, etc) transfer numerically coded data and instructions from conventional media. The operations units perform arithmetic and logical operations on numbers supplied to them from storage. The design of the operations units usually determines whether a computer is serial (numbers handled digit by digit) or parallel (numbers handled as blocks of digits). The design of the operations units, in conjunction with the nature of the storage units, also determines the base of the number system used in the computer. Because many of the components of computers are inherently two-state devices, a base two (binary) number system is often used, particularly in parallel computers. The result is that the input-output units must translate from the conventional decimal system to binary, and vice versa; or, as will be seen, programs can be set up which will accomplish the same function as such machinery. The storage units contain both instructions and data in numerically coded forms. In general, each storage position has an address and at each address one can store instructions or data interchangeably. Much of the flexibility of stored-program computers is due to this feature, because the content at any address can be treated by the operations units. Thus a sequence of instructions can refer to itself and hence alter itself (cf. recursion). The control unit is the heart (or perhaps "brain" would be better) of the stored-program computer. In its simplest form the control unit operates as follows: the instruction address register, a subunit, tells which storage address contains the next instruction. The instruction, which consists of an operation code and (in a single address machine) an address, is decoded. The instruction may specify an operation (operation part) on the contents of some storage address (address part), or it may specify a modification of the instruction address register. If the instruction address register is not modified by the instruction, the address is increased by 1. The control unit then obtains the next instruction thus specified, and so on. If the computer is synchronous, each step is initiated by a central clock which sends synchronized pulses to all parts of the computer; if the computer is asynchronous, each subunit sends out a signal to indicate a completed operation and then goes into a "ready" state.

The chief feature of the stored-program computer, as its name implies, is the stored program. The program consists of a sequence of instructions, each assigned as address in the storage. The possibility of the address part of one instruction referring to the location of another (mentioned in the previous paragraph) permits a program to iterate calculations while changing arguments, parameters, etc. Thus, while some of the instructions in a program carry out actual calculations on data, other instructions modify the program and cause shifts of control. (A shift of control results when the next instruction is <u>not</u> found by adding 1 to the current instruction's address in storage.) The usual procedure in programming is to specify in some very general way the operations to be performed. These operations are then divided into suboperations, subsuboperations, and so on, until one reaches the level of operations available in the computer. (Results like Turing's show that this is possible.) Shifts in control (e.g., when one wants to return to the first instruction of a sequence of instructions just completed) are indicated by a flow diagram. It is possible to program a computer to do any computable operation, and this includes conversion of a number from one base to another. Thus programs can be substituted for actual equipment. Also, an iteration, rather than a table, will often be used to compute the value of a function. Thus, in a sense, one can trade time of calculation (the iteration is usually slower than table look-up) for storage space (the program requires fewer storage locations than the table).

In much the same way that a universal Turing machine can act like any particular Turing machine, a computer can be made to simulate any given complex system (within limits of storage and time). A model is set up by analyzing the complex system into interconnected units. Rules are stated, describing how the inputs to a unit change its state. Storage locations are assigned to each unit, and numbers are used to represent possible states of the units. Program iteration sequences then specify which units (storage locations) act as inputs to any given unit and the way in which the numbers corresponding to the states are to change. Sometimes, instead of carrying out a full simulation, a sampling technique known as the Monte Carlo method is used. The setup is the same as above, but only sample calculations are made (by assigning random numbers to certain states). Just as with axiom systems, simulation requires a good understanding of both the complex system involved and the mechanics of deduction (programming).

6
INFORMATION THEORY

Information theory is a well-developed and extensive field in its own right, but this survey will include only a few of the results most pertinent to automata theory. The results can be divided, roughly, into two categories: those concerned with measure of information and those concerned with channel capacity. The basic definition with respect to the first category is given by the formula

$$H = k \sum_i p_i \log p_i,$$

where H is the quantity of information and p_i is the probability of the i-th state of the system. If k is taken to be 1 and the log is base 2, the quantity of information is given in bits (binary units). (One can readily give, in bits, the storage capacity of computers with two-state storage devices by counting the number of storage positions.) In the case of a Bernoulli process having n outcomes at each trial with the i-th outcome having probability p_i, the measure of information,

H, is given by $H = N \cdot \sum_{i=1}^{n} p_i$, where N is the number of trials. If all the p_i are equal, i.e., $p_i = 1/n$, the measure is maximized, giving H_{max}. (Although the Bernoulli process is discrete, similar theorems can be proved for continuous processes. H_{max} results when the probability distribution is Gaussian and the process has a white or "flat" Fourier power spectrum.) The redundancy, r(x), of a system can be defined in terms of H_{max} as

$$r(x) = 1 - \frac{H(x)}{H_{max}},$$

where H(x) is the information content of the system.

The basic definition of the second category refers to channel capacity. For a noiseless discrete channel, the channel capacity, C, is given by the formula

$$C = \lim_{T \to \infty} \frac{\log_2 N(T)}{T},$$

where N(T) is the number of possible state sequences of duration T. The fundamental theorem for the discrete noiseless channel states that the output of any source can be coded so that the rate of transmission approximates from below, as closely as desired, the channel capacity. This can be achieved by means of a low redundancy code. The coding principle employed is the use of the shorter codes for the higher-frequency sequences. If the channel is noisy, the capacity must be redefined to take account of information lost in the noise. The capacity, C, then is given by

$$C = \max [H(S) - H_R(S)],$$

where H(S) is the channel capacity in the absence of noise, and $H_R(S)$ is the uncertainty about the transmitted signal after the received signal is known (the conditional entropy or information). The fundamental theorem states that the source can be matched to the channel as closely as desired with an arbitrarily small error rate. A new problem that arises in this context is that of error-correcting codes. Parity checks and various generalizations such as the Hamming codes (Ref. 13) provide a way of reducing undetected errors to a rate as small as desired. (At present, parity checks are used on most magnetic tape input-output devices connected to computers. However, the problem of how to introduce redundancy throughout a complex system in order to decrease error probability is an important and relatively untouched problem in automata theory.)

Other portions of information theory (such as theorems on the Fourier transforms of signal frequencies, impulse response, prediction and smoothing, detection of signals in noise) are pertinent, but in some respects they apply more to the engineering of automata than to the theory.

7
NEUROPHYSIOLOGY AND PSYCHOLOGY

The parts of psychology most important to automata theory deal with neurophysiology and theories based upon neurophysiological evidence. The fundamental data concerns the nervous system and its basic component, the nerve cell or neuron. Although there are wide variations in structure, the neuron is generally thought of as consisting of three parts: a cell body or soma, a prominent outgoing fiber or axon which branches profusely before it terminates, and a set of somewhat less prominent branching processes or dendrites. The soma and dendrite bases of each neuron in the central nervous system are closely approached by the axon branches of many other neurons. The points of closest approach are often marked by small swellings called synaptic knobs. As will be explained, these knobs, although not in direct contact with the cell walls of the receiving neuron (there is a separation of a few hundred angstroms) form functional connections from the axon branches to the adjacent cell wall. Such functional connections are called synapses. In the cortex a single neuron may receive synapses from many hundreds of neurons.

Through the use of microelectrodes, Eccles (Ref. 17), more than any other researcher, has contributed an extensive functional analysis of the neuron. The most prominent functional property of the neuron is its threshold behavior. Under "resting" conditions neurons have a voltage drop across the membrane which encloses the cell. If this potential is altered by more than a given amount, the threshold, the neuron emits a pulse of voltage on its outgoing fibers. The amplitude of this pulse is essentially independent of the size of the original disturbance (assuming the disturbance exceeds the threshold).[1] This "firing" of the neuron is followed by a refractory period in which the threshold is first very great and then gradually recovers its normal value. The time required for recovery is of the order of 100 msec. The emitted pulse, however, travels down the axon to the synaptic knobs in a part of a millisecond. According to

[1] It is this last feature which led McCulloch and Pitts and later Kleene (Ref. 22) to investigate logical nets constructed of threshold elements.

current theory, the pulse causes the synaptic knob to secrete chemicals at the synapse, which in turn disturbs the membrane potential of the neurons at which the synapses occur, and so on. There is now a rather elaborate ionic hypothesis, also due to Eccles, which seeks to explain this sequence of events. It should be noted that this is only the main sequence of events and many secondary phenomena occur, some of them quite important. In addition to the various observed secondary phenomena, there is a postulated secondary phenomena which is vital to most neurophysiological learning theories. This is the postulate that if one neuron repeatedly aids in firing another, it becomes better able to do so in the future. Usually this change in "conductance" is considered as localized at the synapse, but there is no firm evidence for this.

Several important system properties of the cerebral nervous system have been revealed by the work of Lashley (Ref. 23) and of Sperry (Ref. 26). They have shown that the cortex, though crisscrossed with cuts through the grey matter (the cerebral region of neuron nuclei), will still function and learn. (The relation between this and the use of redundancy in information theory to obtain error correcting codes should not be overlooked.) Function is significantly disturbed when the cuts penetrate the white matter (a layer of fibers just below the grey matter). Ochs has shown that "U" -fibers exist which dip from the grey matter into the white matter and thence back to the grey matter. Taking into account the work of Lashley and Sperry just mentioned, it may be that the "U" -fibers play a very important role in the integration of neural activity in the cortex. Burns (Ref. 16), working with small neurally isolated slabs of cortex, has shown many interesting macroscopic conduction properties of groups of neurons. In some cases Burns has worked with as few as 5000 to 10,000 functioning neurons. Thus one might hope to simulate some of his experiments with a computer in an attempt to discover which properties of individual neurons are essential to the macroscopic behavior he observes. Only highlights have been presented here; much other important work has been accomplished, especially with EEG apparatus. Only one other set of experiments will be mentioned: the work of Olds (Ref. 14). Olds has shown that there exist certain centers in the midbrain and thalamus which are inherently rewarding or punishing. A rat, for example, will work to obtain electric stimulus through an electrode in the thalamus if the electrode is appropriately placed. These results have had a widespread effect on theories of drive and motivation (persistence of action toward a goal).

Much of the above neurophysiological data has been used by Hebb (Ref. 18) to construct an extremely interesting theory of the formation of concepts in the nervous system. The theory is based upon the neurophysiological learning postulate mentioned above. The central idea of Hebb's theory is the cell assembly. The cell assembly is essentially a group of neurons which are interconnected by synapses of high "conductance" and which act in synchrony. A set of cell assemblies acting on each other in sequential fashion, as determined by the stimulus-response conditions, is called a phase sequence. According to Hebb's theory, the phase sequence corresponds closely to the intuitive idea of a concept of the environmental situation. Taking into account his experimental results, Olds has modified the theory in an interesting way. Each cell assembly is "weighted" according to the number and type of reward-punishment neurons of the thalamus belonging to the assembly. A phase sequence then is considered latent until the sum of the weights of the component cell assemblies exceeds a threshold determined by the needs of the organism. For automata theory, the most important aspects of Hebb's theory and Olds' modifications are the suggestions of how to obtain macroscopic properties of the organism's behavior from the properties and connection schemes of the neurons. Particularly suggestive is the hypothesis of the way the neuro-physiological learning postulate acts through the great number of feedback cycles in the cortex to produce a recognition of macroscopic patterns by a sequential process.

8
SUMMARY AND COMMENTS ON A RESEARCH PROBLEM

In Sec. 2 of this survey a short description was given of the central characteristic of automata theory — the use of abstract deductive theories, based upon the concept of a system being at each moment in one of a given set of states, to study complex natural and artificial systems. Section 3 gave an indication of the way methods, drawn from four established disciplines, are combined to study complex systems. The next four sections elaborated upon the portions of these disciplines particularly important to automata theory. Emphasis was given to the interplay, in automata theory, of methods and problems discussed. This section will be concerned with ways in which the techniques of automata theory might be used in the investigation of a very general class of complex systems.

The systems that will be used as an example are self-organizing systems. That is, the investigation would center on the structural properties which enable a system to adapt (in some sense), by changes in structure, to a range of environments. One way to attack the problem would be to give rigorous definitions to terms like adaptation. The same objective can be attained by giving a rigorous definition of the systems which these properties characterize. A first step in this direction would be the formulation of some very broad class of systems which one feels, intuitively, includes the class of self-organizing systems. More precisely, the procedure would be to set up a formal system having among its interpretations abstract counterparts of self-organizing systems. These abstract counterparts will necessarily be idealizations of the concrete systems; under interpretation, the succession of states in the formal system will predict only approximately or only in certain aspects the succession of conditions in the physical process. The art here is to set up the formal system so that the idealization includes, at least approximately, the features of interest. Once the broader class of systems has been so defined, the next step would be to discover necessary or sufficient conditions for distinguishing the subclass of self-organizing systems. Finally, one would attempt to prove theorems relating structural properties of the self-organizing systems to their adaptability in various environments, and so forth.

The first contribution automata theory could make to this program would be through the concept of growing logical nets (or potentially infinite automata). It seems quite likely that a formal theory of growing logical nets can be set up with the rules of growth actually incorporated in the structure of the net. If, further, the nets are embedded in a 2- or 3-dimensional space, one has an abstract representation of some broad class of systems plus environment. With care, the class of systems so defined should include abstract representatives of the class of self-organizing systems. Now various interpretations or models of the formal system can be simulated. By judicious simulation experiments, it should be possible to discover some of the interpretations which correspond to self-organizing systems. Having done so, the next step would be to use such clues to direct a search for effective normal forms (i. e., algorithms for constructing a normal form for any growing net) which distinguish formal subclasses corresponding to self-organizing systems. There should, in general, be many different types of self-organizing systems. (In fact, there seems no reason to exclude even systems which "evolve" by altering their structure to model the environment, i. e., learning systems.) One such type, of special interest, has its structure, at any future time, only incompletely given at the outset. The system is specified by structural generators ("genes") — the structure at any time being determined by environmental conditions imposed on these generators. The ultimate goal here (using the recursive function approach) would be theorems relating various sets of generators to the range of machines they produce under various restricted environments. (Again simulation should provide useful clues.) Since such systems can be thought of as schema for machines with the blank spaces gradually filled in by the environment, there is a strong relation to the coding of universal Turing machines, but with the coding automatic. For this reason, these systems are of interest not only to genetics, but also to automatic programming and artificial intelligence.

BIBLIOGRAPHY AND REFERENCES

Logic and Metamathematics

Introductory

1. Nagel, E., and Newman, J. R., Gödel's Proof; New York: New York University Press, 1958.
(An excellent account for the general reader of an epoch-making achievement in metamathematics.)

2. Wilder, R. L., Introduction to the Foundations of Mathematics; New York: Wiley, 1952.
(Chapters I and II provide a first-rate introduction to contemporary axiomatics; recursive functions are briefly discussed in Chap. XI.)

Advanced

3. Church, A., Introduction to Mathematical Logic; Princeton: Princeton University Press, 1956, Vol. 1.

4. Davis, M., Computability and Unsolvability; New York: McGraw-Hill 1958.

5. Kleene, S. C., Introduction to Metamathematics; New York: Van Nostrand, 1952.
(Part III contains a detailed development of the theory of recursive functions.)

6. Turing, A. M., "On Computable Numbers, with an Application to the Entscheidungsproblem," Proc. London Math. Soc.; 1936, Vol. 42, ser 2, pp. 230-265.
(This paper is the original account of Turing machines; it is still well worth the reading.)

Stored-Program Computers and Logical Nets

Introductory

7. Kemeny, J. G., "Man Viewed as a Machine", Sci. Am.; April 1955, Vol. 192, pp. 58-67.

8. von Neumann, J., "The General and Logical Theory of Automata," New York: Cerebral Mechanisms in Behavior, ed. Jeffress, L. A.; New York: Wiley. 1951.
(This paper gives an excellent survey of automata in relation to the behavior of living organisms.)

Advanced

9. Burks, A. W., and Wang, H., "The Logic of Automata," J. Assoc. Computing Machinery; 1957, Vol. 4, pp. 193-218 and 279-297.

10. Rochester, N.; Holland, J. H.; Haibt, L. H.; and Duda, W. L., "Tests on a Cell Assembly Theory of the Action of the Brain, Using a Large Digital Computer," I. R. E. Trans. Inform. Theory; 1956, Vol. IT-2, No. 3, pp. S80-S93.

Information Theory
Introductory

11. Bronowski, J., "Science as Foresight," What Is Science?, ed. Newman, J. R.; New York: Simon and Schuster, 1955.
(This paper, besides a discussion of the relation between information and organization, gives a good account of several other subjects related to computers and automata.)

Advanced

12. Shannon, C. E. and Weaver, W., The Mathematical Theory of Communication; Urbana: University of Illinois Press, 1949.

13. Hamming, R. W., "Error Detecting and Error Correcting Codes," Bell System Tech. J.; 1950, Vol. 29, No. 2, pp. 147-160.

Neurophysiology
Introductory

14. Olds, J., "Pleasure Centers in the Brain," Sci. Am.; New York: October 1956, Vol. 195, pp. 105-116.

15. Young, J. Z., Doubt and Certainty in Science, New York: Oxford University Press, 1951.
(The title of this book is somewhat misleading; the book describes with great clarity the brain processes which produce the higher activities of man.)

Advanced

16. Burns, B. D., The Mammalian Cerebral Cortex; Baltimore: Williams and Wilkins, 1958.
(This book contains accounts of Burns' important experiments with isolated slabs of living cerebral cortex.)

17. Eccles, J. C., The Physiology of Nerve Cells; Baltimore: Johns Hopkins Press, 1957.

18. Hebb, D. O., The Organization of Behavior; New York: Wiley, 1949.
(Hebb's book offers suggestive hypotheses and penetrating insights as to the way in which behavior arises out of the interaction of the neurons in the central nervous system.)

19. Sholl, D. A., The Organization of the Cerebral Cortex; New York: Wiley, 1956.

Other Papers Cited in the Survey

20. Burks, A. W., and Wright, J. B., "Theory of Logical Nets," Proc. I. R. E.; 1953, Vol. 41, pp. 1357-1365.

21. Chomsky, N., "Three Models for the Description of Language," I. R. E. Trans. Inform. Theory; 1956, Vol. IT-2, No. 3, pp. S113-124.

22. Kleene, S. C., "Representation of Events in Nerve Nets and Finite Automata," pp. 3-41 in Automata Studies, ed. Shannon, C. E., and McCarthy, J., Princeton: Princeton University Press, 1956.

23. Lashley, K. S., "Transcortical Association in Maze Learning," J. Comp. Neurol.; 1944, Vol. 80, pp. 257-281.

24. Ochs, S., "The Direct Cortical Response," J. Neurophysiol.; 1956, Vol. 19, pp. 513-523.

25. Shannon, C. E., "A Universal Turing Machine with Two Internal States," pp. 157-165 in Automata Studies, ed. Shannon, C. E., and McCarthy, J., Princeton: Princeton University Press, 1956.

26. Sperry, R. W., "Cerebral Regulation of Motor Coordination in Monkeys Following Transection of Sensori-Motor Cortex," J. Neurphysiol.; 1947, Vol. 10, pp. 275-293.

from: Proceedings of the 1960 Western Joint Computer Conference

259
9.2

ITERATIVE CIRCUIT COMPUTERS

John H. Holland

The University of Michigan
Ann Arbor, Michigan

SUMMARY

The paper first discusses an example of a computer, intended as a prototype of a practical computer, having an iterative structure and capable of processing arbitrarily many words of stored data at the same time, each by a different sub-program if desired. Next a mathematical characterization is given of a broad class of computers satisfying the conditions just stated. Finally the characterization is related to a program aimed at establishing a theory of adaptive systems via the concept of automaton generators.

I

INTRODUCTION

Computers constructed of hundreds of millions of logic and storage elements will require an organization radically different from present computers if the elements are to be used efficiently in computation. It should be possible to process arbitrarily many words of stored data at the same time, each by a different sub-program if desired. In addition the structure of the computer should be iterative or modular in order to allow efficient use of template techniques in its construction.

The present paper contains a mathematical characterization of a broad class of computers satisfying these conditions. This class, with appropriate interpretation of the symbols, includes representatives structurally and behaviorally equivalent to each of the following types of automata:

1) Turing machines (with 1 or more tapes)[12,9]
2) Tessellation automata (Von Neumann, Moore)[14,7]
3) Growing logical nets (Burks-Wang)[2,1]
4) Potentially-infinite automata (Church)[3]

The class also contains automata which, in various senses, are generalizations of each of these four types.

Ultimately, for the designer, the value of such a characterization depends upon whether or not it can actually suggest designs for solid-state computer. Section II of this paper discusses a possible abstract prototype for such a computer--it is one of many alternatives which can be defined and investigated with the help of the characterization. (A similar computer is discussed in greater detail in the Proceedings of the 1959 Eastern Joint Computer Conference)[5]. Section III summarizes the mathematical characterization and considers its interpretation. Section IV relates this paper to a program (begun by the author in 1958) which has as its objective a theory of adaptive systems.

II

AN ITERATIVE CIRCUIT COMPUTER

The computer outlined in this section is presented primarily to suggest something of the class of computers included in the characterization summarized in the next section. At the same time, however, the order code, addressing schemes, etc. where chosen to reflect their counterparts in present computers. In this sense, the computer can also be thought of as a prototype of a practical computer--assuming that the large numbers of components required can be provided economically. Because the computer can execute an arbitrary number of sub-programs simultaneously, and because the sub-programs are spatially organized, its operation is of course considerably different from present computers.

The computer can be considered to be composed of modules arranged in a 2-dimensional rectangular grid; the computer is homogeneous (or iterative) in the sense that each of the modules can be represented by the same fixed logical network. The modules are synchronously timed and time for the computer can be considered as occurring in discrete steps, $t = 0, 1, 2, \ldots$.

Basically each module consists of a binary storage register together with associated circuitry and some auxiliary registers. At each time-step a module may be either active or inactive. An active module, in effect, interprets the number in its storage register as an instruction and proceeds to execute it. There is no restriction (other than the size of the computer) on the number of active modules at any given time. Ordinarily if a module $M(i,j)$ at coordinates (i,j) is active at time-step t, then at time-step $t+1$, $M(i,j)$ returns to inactive status and its successor, one of the four neighbors $M(i+1,j)$, $M(i,j+1)$, $M(i-1,j)$, or $M(i,j-1)$, becomes active. (The exceptions to this rule occur when the instruction in the storage register of the active module specifies a different course of action as, for example, when the instruction is the equivalent of a transfer

260
9.2

instruction).

The successor is specified by bits s_1, s_2 in M(i,j)'s storage register. If we define the line of successors of a given module as the module itself, its successor, the successor of the successor, etc., then a given sub-program in the computer will usually consist of the line of successors of some module. Since several modules can be active at the same time the computer can in fact execute several sub-programs at once. We have noted parenthetically that there are orders which control the course of action--there are also orders equivalent to store orders which can alter the number (and hence the instruction) in a storage register. Therefore, the number of sub-programs being executed can be varied with time, and the variation can be controlled by one or more sub-programs.

The action of a module during each time-step can be diveded into three successive phases:

1) During phase one, the initial phase of each time-step, a module's storage register can be set to any arbitrarily chosen value and its auxiliary registers to any desired condition. The numbers and conditions thus supplied are the computer's input. Although the number in the storage register can be arbitrarily changed at the beginning of each time-step, it need not be; for many purposes the majority of modules will receive input only during the first few moments of time ("storing the program") or only at selected times t_1, t_2, . . . ("data input"). Of course, some modules may have a new number for input at each time-step; in this case the modules play a role similar to the inputs to a sequential circuit.

2) During phase two, an active module determines the location of its operand set, the set of storage registers upon which its instruction is to operate. This the module does by, in effect, opening a branching path (sequence of gates) to the operands. The path-building action depends upon two properties of modules:

First, by setting bit p in its storage register equal to 1, a module may be given special status which marks it as a point of origination for paths; the module is then called a P-module.

Secondly, each module has a neighbor, distinct from its successor, designated as its predecessor by bits q_1, q_2 in its storage register; the line of predecessors of a given module M_0 is then defined as the sequence of all modules $[M_0, M_1, \ldots, M_k, \ldots]$ such that, for each k, M_k is the predecessor of M_{k-1} and M_{k-1} is the successor of M_k. Note that the line of predecessors may in extreme cases be infinitely long or non-existent. The line of predecessors of an active module ordinarily serves to link it with a P-module (through a series of open gates). During the initial part of phase two the path specification bits $y_0, \ldots, y_n$ and $d_0, \ldots, d_3$ in the storage register of an active module M_0, are gated down its line of predecessors to the nearest P-module (if any) along that line. The path specification bits are then used by the P-module to open a branching path to the operand set of the active module.

Each path must originate at a P-module. The modules belonging to a given path can be separated into sub-sequences call segments. Each segment consists of y modules extending parallel to one of the axes from some position (i,j) through positions $(i+b_1,\ j+b_2)$, $(i+2b_1,\ j+2b_2)$, . . . , $(i+(y-1)b_1,\ j+(y-1)b_2)$, where $b_1 = \pm 1$ or 0 and $b_2 = \pm(1-b_1)$; the module at $(i+yb_1,\ j+yb_2)$ will be called the termination of the segment. Each module possesses four *-registers and if the module belongs to a segment in direction (b_1, b_2) the appropriate *-register, $(b_1, b_2)^*$, is turned on gating lines between (i,j) and $(i+b_1,\ j+b_2)$. Since each *-register gates a separate set of lines, a module may (with certain exceptions) belong to as many as four paths. Once a *-register is turned on it stays on until it is turned off; thus a path segment, once marked, persists until "erased".

Each segment of a path results from the complete phase two action of a single active module; however, since a path may branch, more than one segment may result in one time-step from the action of a given active module. After the digits $y_n, \ldots, y_0$, $d_3, \ldots, d_0$ are gated to the nearest P-module along the line of predecessors of the active module, new segments are constructed at the termination of each branch of the path originating at the P-module. Note that, because of the branching, there will be more than one path termination.

Branching is controlled by the digits $d_3, \ldots, d_0$. To each of the four digits $d_3, \ldots, d_0$ corresponds one of the four neighbors at each branch termination. If $d_i=1$ then, when the path is extended, at each existing path termination a new branch will be sent through the ith neighbor parallel to the axis.

Path extension takes place only when bit $y_n=0$; then bits $y_{n-1}, \ldots, y_0$ determine the common length of the new segments and bits $d_3, \ldots, d_0$ determine their directions. If $y_n=1$ then final path segments, if any, in the directions specified by $d_3, \ldots, d_0$ are erased (bits $y_{n-1}, \ldots, y_0$ not being used in this case). In order to prevent interference of one path with another, or with itself, a set of priority and interlock rules are required. These rules will not be specified here but the interested reader can see a complete set of such rules for a similar computer in the 1959 E.J.C.C. paper cited previously.[5]

3) During phase three, an active module executes the instruction contained in its storage register. This involves the following modules: the active module itself holds the order code in bits i_2, i_1, i_0 of its storage register; the storage registers of the modules terminating the nearest path contain the set of words to be operated on (the operand set); finally there must be a module which serves as arithmetic unit. In order to serve as an arithmetic unit, bits (p,a) in the storage register of a module must first be set to the value (0,1), giving the module special status--A-module status. (Note that this means a module in P-module status, p=1, cannot be an A-module). If

M(i,j) is an active module then the first A-module along its line of predecessors serves as the arithmetic unit.

A short, though representative, set of orders follows:

(i) Execution of OR/ADD causes the following sequence of actions: first the numbers stored at the modules in the operand set are transferred down the branches of the path toward the P-module; as these numbers meet at branch-points a resultant is formed equal to the bit-by-bit disjunction of the incoming numbers (i.e. bit j in the resultant is 1 only if at least one of the incoming numbers has 1 at position j); when the final resultant is formed at the P-module it is transferred along the line of predecessors to the nearest A-module; there the number is added to whatever number is in the storage register of the A-module. Note that this sequence of actions takes place wholly within phase three of the time-step in which the instruction is executed.

(ii) Execution of AND/ADD proceeds just as OR/AND except that a bit-by-bit conjunction (output bit is 1 only if all corresponding input bits are 1) takes the place of bit-by-bit disjunction.

(iii) Execution of STORE causes the number in the storage register of the nearest A-module to be transferred to the storage registers of all modules in the operand set.

(iv) Execution of TRANSFER ON MINUS depends upon the number in the storage register of the nearest A-module. If, in this number, $y_n=0$ then at the end of phase three the active module becomes inactive and its successor becomes active. If $y_n=1$ then each of the modules in the operand set, rather than the successor, become active.

(v) NO ORDER causes the execution phase to pass without the execution of an order.

(vi) STOP causes the active module to become inactive without passing activity on to its successor at the next time-step.

With the exception of the TRANSFER and STOP orders, the active module becomes inactive and its successor becomes active at the conclusion of phase three. Just as in the case of phase two some rules are required to prevent interference of active modules and to provide for cases where there is no nearest A- or P-module along the line of predecessors (the reader is again referred to the 1959 E.J.C.C. paper)[5].

The storage register of each module in the present formulation consists of n+14 bits labelled in the following order:

bit number:	n+14	n+13	...	14	13	12	11	10	9	8	7	6	5	4	3	2	1
label:	y_n	y_{n-1}	...	y_0	d_3	d_2	d_1	d_0	i_2	i_1	i_0	s_1	s_2	q_1	q_2	p	a

The function, in the active module, of each bit group has already been discussed.

III

MATHEMATICAL CHARACTERIZATION OF ITERATIVE CIRCUIT COMPUTERS

One purpose of the mathematical characterization summarized here is to define the class of iterative circuit computers precisely enough to allow mathematical deduction to be used in their study. This property of the characterization will be used in later work in an attempt at establishing a theory of adaptive systems (see the next section). At the same time the characterization can be used to generate a wide range of computer prototypes, each with different structural and operational characteristics. Thus, the characterization can also be of help in the design of solid-state computers.

The characterization is made up of the following parts:

1) The positions of the modules are indexed by the elements of a finitely-generated abelian group, A. The particular group chosen determines the "geometry" of the network; for instance, by choosing the appropriate group, the modules can be arranged in a plane, or a torus, or an n-dimensional cube, etc. Thus, for a computer with the modules arranged in a 2-dimensional rectangular grid 1000 modules on a side, A would be the abelian group with two generators a_1, a_2 satisfying the relations

$$1000\ a_1 = e$$
$$1000\ a_2 = e$$

where e is the identity element of the group.

The group, A, is restricted to being a finitely-generated abelian group for several technical reasons. One reason is that the elementary theory of such groups is decidable. When taken with the rest of the definition of iterative circuit computers, this implies that the operation of the computer is effectively defined. Also any such group can be decomposed into a direct product of cyclic subgroups. Thus the elements of the group can be represented uniquely as n-tuples on the basis of certain sets of generators of the group (in other words, the modules are arranged in a "regular" fashion).

2) In order to determine the immediate neighbors of a module we must specify a finite set, $A^\circ = (a_1, \ldots, a_k)$, of elements selected from the

group A. Then the set of immediate neighbors of the module at $\alpha \in A$ are the modules at $a_i(\alpha) = \alpha + a_i$ for all $a_i \in A^*$, where $+$ is the group operation. For example, if we have a module at coordinates (i, j) relative to generators a_1, a_2 and we wish its immediate neighbors to be at coordinates $(i+1, j)$, $(i, j+1)$, $(i-1, j)$, and $(i, j-1)$ then we could choose $A^* = \{a_1, a_2, a_1^{-1}, a_2^{-1}\}$, where a_i^{-1} is the group inverse of a_i.

3) The state of _each_ module in the computer at each time t must be drawn from a finite set, S, of allowable states. $S = X \times Y$, the cartesian product of the sets X and Y. X can be any finite set of elements--the elements will be called "storage states"; $Y = \prod_{i=1}^{k} Y_i$, a cartesian product of the sets $Y_i = R = \{a_1, \emptyset\} \square \ldots \ldots \square \{a_k, \emptyset\}$.

In what follows the notation S_α^t will be used to denote the state of the module at position α at time t. A similar convention will be used for the components of S. Note that the elements of Y can be though of as k by k matrices. The matrix Y_α^t which holds for the module α at time t is called the connection matrix of α at time t. The ith row of Y_α^t, $Y_{\alpha i}^t = (Y_{\alpha i 1}^t, Y_{\alpha i 2}^t, \ldots, Y_{\alpha i k}^t)$, specifies which of the k immediate neighbors of α are connected through α to the module at $a_i^{-1}(\alpha)$; if $Y_{\alpha i j}^t = a_j$ then the module at $a_j(\alpha)$ is connected through α to the module $a_i^{-1}(\alpha)$ otherwise not.

4) Changes in the k rows $Y_1, \ldots, Y_k$ of the connection matrix Y from one time step to the next, $Y_{\alpha i}^t$ to $Y_{\alpha i}^{t+1}$, are determined by a set of k projections:

$$P_i : S \rightarrow R = \{a_1, \emptyset\} \square \ldots \square \{a_k, \emptyset\}.$$

The way in which these changes are effected will be described in terms of the transition equations to be given shortly.

5) Change in the storage state from one time step to the next, X_α^t to X_α^{t+1}, is determined by a function

$$f : \prod_{i=1}^{k} S \rightarrow X$$

which will be called the "sub-transition function". Again f can be best explained in terms of the transition equations.

A particular selection for each of the five parts described in 1) through 5), $(A, A^*, X, \{P_i\}, f)$, determines a particular iterative circuit computer.

In addition a function $B : \{A\} \square t) \rightarrow S$ may be effectively defined for some pairs (α, t); B_α^t gives the input to α at time t, when B is defined for (α, t). Once a particular iterative circuit computer is specified (and its initial state is given), the transition equations together with the function B determine the operation of the computer.

The transition equation for the connection matrix Y_α^t in terms of its elements, $Y_{\alpha i j}^t$, is:

$$Y_{\alpha i j}^{t+1} = \emptyset \quad , \text{ if } P_{\alpha i j}^t = \emptyset \qquad \text{("erasure")}$$

$$= Y_{\alpha i j}^t \quad , \text{ if } P_{\alpha i j}^t = a_j \text{ and } Q_{\alpha i j}^t = 0$$
$$\text{("no change")}$$

$$= P_{\alpha i j}^t \quad , \text{ if } P_{\alpha i j}^t = a_j \text{ and } Q_{\alpha i j}^t = 1$$
$$\text{("construction")}$$

where $P_{\alpha i j}^t$ is the jth component of $P_i(S_\alpha^t)$

and $\quad Q_{\alpha i j}^t = \&(q_{a_j(\alpha)j1}^t, \ldots, q_{a_j(\alpha)jk}^t)$

$$q_{\beta i j}^t = 0 \quad , \text{ if } Y_{\beta i j}^t = \emptyset \text{ and } P_{\beta i j}^t = a_j$$

$$= 1 \quad , \text{ if } P_{\beta i j}^t = \emptyset$$

$$= \&(q_{a_j(\beta)j1}^t, \ldots, q_{a_j(\beta)jk}^t),$$
$$\text{otherwise}$$

defining $\&(c_1, \ldots, c_k) = 1$, if all $c_j = 1$
$$= 0 \text{ , otherwise.}$$

As mentioned in 3) above, row i of the connection matrix Y_α^t can be interpreted as specifying a set of modules $a_j(\alpha)$ connected through α to $a_i^{-1}(\alpha)$; for each such j, row j of the connection matrix $Y_{a_j(\alpha)}$ may specify other modules $a_h a_j(\alpha)$ connected, via $a_j(\alpha)$ and then α, to $a_i^{-1}(\alpha)$; appropriate rows of the matrices $Y_{a_h a_j(\alpha)}$ may specify still others; etc. In other words the matrix Y_β^t tells how information is to be channeled through β to its immediate neighbors, the matrices for these neighbors tell how the information is to be sent on from there, etc. In this way each module serves as the base of what may be a complex branching tree channeling information to it. It will be seen (in the transition equations for X_α^t) that modules belonging to the tree for α pass information to α _without_ a time-step delay.

The transition equation for Y_α^t can now be given the following interpretation: Broadly, $P_{\alpha i}^t$ (in $P_i(S_\alpha^t)$) specifies changes in row i of the connection matrix Y_α^t while $Q_{\alpha i}^t = (Q_{\alpha i 1}^t, \ldots, Q_{\alpha i k}^t)$ prohibits certain of these changes. More specifically,

$Q^t_{\alpha i j} = 0$ prohibits construction of a new connection from $a_j(\alpha)$ through α to $a_i^{-1}(\alpha)$ at time t. $Q^t_{\alpha i j} = 0$ just in case construction of a new connection is indicated somewhere in the tree for α. Some thought will show that this rule (others, at least superficially more general, could have been chosen) implies the following desirable conditions:

(i) a cycle of connections without delay cannot be formed (operation of modules belonging to such a cycle would in general be indeterminant--consider the analagous case of a set of one-way, non-delay switches arranged in a cycle)

(ii) the tree for any given module α never includes more than a finite number of modules (even if, for theoretical purposes, the group A is infinite).

The transition equation for X^t_α is:

$$X^{t+1}_\alpha = f\,(S'_{a_1(\alpha),1}(t),\ S'_{a_2(\alpha),2}(t),\ \ldots,\ S'_{a_k(\alpha),k}(t))$$

$$S'_{\beta,i}(t) = S^t_\beta,\ \text{if } Y^{t+1}_{\beta i} = (\emptyset,\ldots,\emptyset)$$
$$\text{and } B^{t+1}_\beta \text{ not defined}$$

$$= B^{t+1}_\beta,\ \text{if } Y^{t+1}_{\beta i} = (\emptyset,\ldots,\emptyset)$$
$$\text{and } B^{t+1}_\beta \in S$$

$$= f\,(S'_{Y_{\beta i 1}(\beta),1}(t),\ \ldots,\ S'_{Y_{\beta i k}(\beta),k}(t)) = Y^{t+1}_\beta,$$
$$\text{if } Y^{t+1}_{\beta i} \neq (\emptyset,\ldots,\emptyset)$$

If $Y^{t+1}_{\beta i j} = \emptyset$ then define $S'_{\emptyset(\beta),j}$

$$= S_{a_j(\beta)},\ \text{if } B^{t+1}_{a_j(\beta)} \text{ not defined}$$

$$= B^{t+1}_{a_j(\beta)},\ \text{otherwise.}$$

Under interpretation the transition equation for X^t_α specifies the storage state, X^{t+1}_α, in terms of the states at time t, S^t_β, of the modules β belonging to the connection tree for α. Note that, because of the recursive definition of S'(t), f may be iterated several times in the determination X^{t+1}_α--compare this to the determination of the output of a tree of switches without delays.

IV

TOWARD A THEORY OF ADAPTIVE SYSTEMS

As already mentioned, the work reported here is part of a larger effort which has as its goal a theory of adaptive systems. The effort is an individual one and, of course, reflects particular biases of the author. This section will discuss the relation of the present paper to the broader program.

The first step of this program was the description of a computer which could simulate the operation and, in certain respects, the structure of any automaton (growing or fixed). The second step, summarized here in part III, consisted in giving a general and formal description of computers like the one first obtained--the iterative circuit computers. The resulting mathematical characterization represents a broad class of machines, of arbitrary geometries, etc.; by an appropriate choice of $(A, A^\circ, X, \{P_i\}, f)$ it is possible to represent directly not only the behavior but also the changing structure and local operation of any given potentially-infinite automaton, tessellation automaton, n-tape Turing machine, or growing logical net. (In this respect note, for instance, that the class of iterative circuit computers properly contains A. Church's class of potentially-infinite automata--any two modules in an iterative circuit computer may eventually become connected so that either affects the other in a single time-step, whereas in a potentially-infinite automaton the corresponding delay, in time-steps, increases with increasing separation and is a constant for any given separation).[3]

For an iterative circuit computer the ideas of sub-program and automaton are, in an important sense, interchangeable. For any automaton, a sub-program can be written which not only has the same behavior but also the same changing structure and local operation. On the other hand, a given sub-program will in general occupy a finite number of modules in the computer and will have its action (or state) determined by bordering modules and the input function B. Thus, as a survey of the characterizing equations will show, an automaton or growing logical net can be constructed which mimics the sub-program.

It is important to note that sub-programs can be set up which, for instance, can shift themselves from one set of modules to another set, i.e. from one position to another. Thus the underlying geometry of the iterative circuit computer (given by A and A°) in effect determines the geometry of a space in which the sub-program is embedded; the transition equations then serve, in a sense, as the laws of this universe. For this reason the sub-programs of a given iterative circuit computer will often be spoken of as "embedded automata". This view of the results of the second step leads directly to the third step.

264
9.2

The central object of the third step is to provide formal apparatus for the implicit definition of automata--definition by means of generators and relations on these generators. Implicit definition of an automaton is analogous to the implicit definition of an algebraic group. In the case of the group, instead of giving an explicit listing of the elements of the group and their interrelations, the group is defined (often quite compactly) in terms of a set of generating elements and relations on products of the generators. In a similar sense an automaton can be implicitly specified by an initial set of elements and a set of growth rules.

The mathematical characterization of iterative circuit computers provides the apparatus needed for a precise formulation of automaton generators. Specifically, the generators will be sub-programs which can be thought of as (relatively) elementary embedded automata having the following properties:

(i) movement--the generators will in general be capable of shifting as a unit from one position to another (as specified by input $B(\alpha,t)$ or the state of adjacent modules--note that motion may be random if the input sequences $B(\alpha,t)$ are random),

(ii) connection--generators will combine under conditions specified internally (within the sub-programs) to form larger sub-programs capable of moving and acting as units,

(iii) production--generators can alter the state of adjacent modules (note that a sufficiently complicated generator could directly duplicate itself).

The generators will act upon other generators or other sub-programs present in the computer ("precursors") by connecting them or breaking them into components. The generators will all be acting simultaneously and if a given generator duplicates itself the duplicate will also in general be active.

So that any possible automaton can be defined in terms of the generators it is necessary to choose the set of generators so that any possible program for the computer can be represented by an appropriate connected set of generators (cf. the process of picking a set of instructions sufficient for a universal computer). Once this is done, the generation of particular automata can be effected by controlling rates of production and connection, movement and contact, the nature of precursors present, etc. That is, the relations on the generators will consist of specifying the initial states and input sequences which control such factors. For any given iterative circuit computer and set of generators, the relations possible can be given an appropriate equational form. Under one approach, the way in which the generators are initially connected and the nature of the precursors in the "environment" are sufficient together to specify the generated automaton. Things become particularly simple if the generators cannot interpenetrate when moving (a kind of "billiard ball physics").

A given program, because of the loops or iterations, is much more compact than the complete sequence of steps in the calculations it controls. In the same sense the connected system of generators which specify a given automaton will be much more compact than the automaton generated. Thus the automaton can have scattered throughout its structure complete implicit descriptions of its structure--such considerations play an important part in the study of self-repairing automata.

In the implicit definition of an automaton certain feedback phenomena play a crucial role. The feedback phenomena can be to some extent isolated by observing at what level a given generator system falls in the following succession of categories (each of which properly includes its successor):

(i) productive systems--the generator system produces other generators or precursors,

(ii) autocatalytic systems--the generator system produces generators or precursors which are used in its construction, i.e. the system produces some of its own components,

(iii) self-duplicating systems--the generator system produces duplicates of itself.

Such considerations lead directly from step three to step four and from work in progress to work which lies in the future.

The central object of step four will be to define the term "adaptive system" for embedded automata. Because of the formal nature of the definition, it then becomes possible to investigate these adaptive systems deductively (and by simulation). The beginnings of such a definition lie in the following consideration: With the help of concepts such as autocatalytic and self-duplicating generator systems it is possible to define such concepts as steady-state equilibria and homeostasis for embedded automata. In fact one can go quite far in this direction obtaining discrete state relaxation processes (Southwell), morphogen standing-wave phenomena (Turing) and so forth.[11, 13] Automata exhibiting these properties will usually have the desirable property that small changes in structure result in small changes in behavior (at least over a certain range). Thus the behavior of a given automaton of this type gives some indication of the behavior of all automata of similar structure ("hill-climbing" techniques become applicable). If the generator system for such an automaton has a hierarchical structure, then a small change in structure produces a small effect in proportion to the "position" of the change in the hierarchy. That is, a generator system may consist of autocatalytic or homeostatic systems, systems of these (which may or may not be autocatalytic or homeostatic), etc.; changes at the upper levels of the hierarchy will generally have a greater effect than those at lower levels. By making changes first at the highest level and then at progressively lower levels of the hierarchy, it should be possible to narrow down rather quickly to any automaton in this category having some initially prescribed behavior.

Changes in the generated structure result when relations on the generators are altered. The effect of such alterations can perhaps be more clearly seen under the following interpretation. The generation of a particular automaton can be looked at as if all possible generation processes are going on simultaneously but at different rates. Some will be going very slowly (infinitely slowly in the limit) while others will be proceeding very rapidly. The way in which these rates are changed in order to change the generated automaton will have important consequences with respect to the adaptiveness of the overall system (cf. A. L. Samuel's work on changing the weights of a "checker-move tree").[10]

As a final point it should be noted that the environment of an embedded automaton can be made as simple or complex as desired. Since adaptation must be defined in terms of the range of environments in which the automaton is to be embedded, this is an important factor. It has already been noted that the environment may contain other sub-programs (precursors) or in fact other embedded automata. The latter case amounts to an implicit definition of the environment since only the initial state and internal rules of each of the embedded automata need be given. Contrast this with an explicit definition which would require a point-by-point, time-step-by-time-step description of the state of the environment. If the precursors in the environment are relatively elaborate and sophisticated then the adaptation process will look similar to a heuristic learning system (cf. Newell-Shaw-Simon)[8] If precursors are absent or simply generators then the adaptation process will look more like the processes considered by Friedberg.[4] It seems likely that implicit definition of the environment will play an important part in the development of step four.

BIBLIOGRAPHY

1. Burks, Arthur W., "Computation, Behavior, and Structure in Fixed and Growing Automata" University of Michigan Technical Report ONR Contract 1224(21) (1959).
2. Burks, Arthur W. and Hao Wang, "The Logic of Automata" J. Assoc. Computing Mach. 4, 2 & 3, 193-218, 279-297 (1957).
3. Church, Alonzo, "Application of Recursive Arithmatic in the Theory of Computers and Automata" notes from summer conference course in Advanced Theory of the Logical Design of Digital Computers, The University of Michigan (1958).
4. Friedberg, R.M., "A Learning Machine: Part 1" IBM Journal of Research and Development 2, 1,2-13 (1958).
5. Holland, J.H., "A Universal Computer Capable of Executing an Arbitrary Number of Sub-Programs Simultaneously" Proc. 1959 Eastern Joint Computer Conference.
6. Kleene, S. C., "Representation of Events in Nerve Nets and Finite Automata" in Automata Studies Annals of Mathematics Studies no.34, Princeton (1956).
7. Moore, Edward F., "Machine Models of Self-Reproduction" paper (560-52) at October Meeting of the American Mathematical Society, Cambridge, Mass. (1959).
8. Newell, A., J.C. Shaw and H.A. Simon, "Empirical Explorations of the Logic Theory Machine: A Case Study in Heuristics", Report P-951, Rand Corporation (1957).
9. Rabin, M.O., and D. Scott, "Finite Automata and Their Decision Problems" IBM Journal of Research and Development 3, 2,114-125 (1959).
10. Samuel, A.L., "Some Studies in Machine Learning, Using the Game of Checkers" IBM Journal of Research and Development 3, 3,210-229 (1959).
11. Southwell, R.V., Relaxation Methods in Engineering Science; a Treatise on Approximate Computation, Clarendon Press, Oxford (1940).
12. Turing, A.M., "On Computable Numbers, with an Application to the Entscheidungsproblem" Proc. Lond. Math. Soc. (2), 43, 230-265 (1936).
13. Turing, A.M., "The Chemical Basis of Morphogenesis", Phil. Trans. Roy. Soc., ser. B, 237, 37 ff. (1952).
14. Von Neumann, J., The Theory of Automata, unpublished manuscript.

* * * *

This research was supported by the U.S. Army Signal Corps through contract OA-36-039-SC-78057, and by the National Science Foundation through grants G-4790 and G-11046. The paper was written while the author was a member of the Institute of Science and Technology at the University of Michigan.

CONCERNING EFFICIENT ADAPTIVE SYSTEMS

John H. Holland
Communications Sciences Laboratory
University of Michigan, Ann Arbor, Michigan

> "It is interesting to . . . reflect that these elaborately constructed forms, so different from each other, and dependent upon each other in so complex a manner, have all been produced by laws acting around us. These laws, taken in the largest sense, being Growth with Reproduction; Inheritance . . .; a Ratio of Increase so high as to lead to a Struggle for Life, and as a consequence to Natural Selection, entailing Divergence of Character and the Extinction of less-improved forms."
>
> --Darwin

1. Introduction

This study of adaptive efficiency is based upon the framework described in "Outline for a Logical Theory of Adaptive Systems" [5]. Its principal object is to give some idea of the methods appropriate to the framework. Two considerations prompted the choice of "adaptive efficiency" for this purpose:

(1) the study of efficiency has lead to important results in related areas such as information theory;

(2) perliminary work indicated that, along this line, statistical mechanics applies in a direct and simple way to the generation procedures and generated systems of automata theory.

The results are derived in a narrow and elementary context, but the methods used (and the results) appear to extend naturally to broader contexts.

Section 2 ('General discussion') gives a general statement of the observations motivating the paper. The translation of these observations to the more rigorous framework provided by automata theory requires first of all an appropriate class of automata; the class of automata used has been described elsewhere [6] and only a few relevant points will be reviewed in section 3 ('Abstract basis'). Section 4 ('Adaptive efficiency') considers the concept of efficiency as it applies to a restricted subclass of the systems described in section 3. It is shown that acquisition of new information generally forces a concurrent. less efficient use of information already accumulated. The main result gives the rate of acquisition of information which (as a function of the concurrent loss of efficiency) produces optimal adaptation. Loosely, the result shows both that no system of the type considered should consign more than half of its effort to acquisition of new information, and that there are reasonable conditions under which this limit should be closely approached. The final section ('Summary') discusses the particular results and their generalization to the full range of systems described in section 3.

2. General discussion

Classically, adaptation is a process whereby an organism is modified to fit it (or its progeny) more perfectly for existence in its environment. This classical view, if one pays it close heed, sets distinctive conditions on a theory of adaptive systems: The statement emphasizes that it is not sufficient to characterize just the system's internal processes. The environment (or range of possible environments), the information received therefrom and the ways the system can affect the environment must also be characterized. And, when this is done, the phrase ". . . to fit it more perfectly for existence. . ." still requires attention. The phrase suggests both a condition and a way of meeting it: Clearly, the theory should provide a formal means of determining which of two organizations is the fittest in a given environment. That is, the theory should include a fitness-measuring function, either directly or indirectly defined, enabling comparison of systems and selection of the fittest. Only then is it possible to discuss the *process* of adaptation within the context of the theory. ". . . to fit it . . . for existence . . ." suggests that the fitness measure be defined indirectly in terms of a survival criterion. As a matter of fact the problem of fitness can quite generally be rephrased in terms of survival under conditions imposed by the environment [4]. The result is a differential selection of systems according to fitness—the fittest systems at any point in time being those which have persisted. In general terms, then, the core of a theory of adaptive systems should be a study of differential selection.

A living organism constantly exchanges material and energy with its environment—an environment which, even when it is homogeneous in the large, involves important local variations.

(One can observe similar interactions in any adaptive system, natural or artificial.) If a species of organism is to survive it must at least partially control this interaction. Because the environment does vary, this control can only be apropos if the organism receives and employs information about its environment. A system receiving no information from its environment obviously cannot base its control procedures upon the condition of the environment; "complete" information, on the other hand, opens the possibility of perfect control. Moreover, moving from the first extreme of no information to the second of complete information, one would expect a steady improvement in control possibilities. If the information is received at a rate r, then adaptation is limited accordingly; a system can alter its organization no more rapidly than it receives the relevant information. Thus, the efficiency of a system's control is limited by the rate at which it receives relevant information from the environment. And, of two similar systems in similar environments, the one exercising more efficient control will be the one favored by differential selection. Or so it seems intuitively.

It is another matter to give precision to this argument. The argument as presented has, in common with most heuristic arguments, a very loose texture. It can hardly be validated or invalidated in its given form and it is almost impossible to ascertain what elements of truth lie within it. In particular, although a relation between adaptive efficiency and information input rate is suggested, the precise nature of the relation and its usefulness can only be evaluated in a more rigorous context. The remainder of the paper will be devoted to a deeper look at this relation.

3. Abstract basis

This paper makes contact with automata theory through the formally defined class of iterative circuit computers. As the introduction notes, an earlier paper [5] describes the use of this class as a basis for studies of adaptation. Only points relevant to the present inquiry will be touched upon in this section.

In general terms, the study of adaptation is a study of how systems generate methods enabling them to adjust efficiently to their environments. Let us equate "method" with "program of a specified universal computer." From a more abstract viewpoint, then, an adaptive system is a schema for generating programs in accordance with the dictates of the environment. Let any such schema be called a *generation procedure*. The population of programs generated at a given time can be considered the repertory of methods available to the adaptive system at that time. The discussion here will be based upon generation procedures defined in terms of a selected finite set of programs, called *generators*, and a graph, called a *generation tree:*

(1) No restriction is to be placed upon the programs selected for the set of generators; in one case the set of generators may include just the equivalents of individual instructions, in another it may include highly sophisticated heuristic programs. If some generation procedure is supposed to generate all programs of a specified universal computer, the set of generators must include a complete set of instructions for that computer.

(2) The generation tree specifies the combining processes whereby the programs are formed from the generators. The vertices of the tree can be divided into two categories: major vertices and auxiliary vertices. Each major vertex of the graph represents a distinct program which can be formed from the generators. Each auxiliary vertex indicates a distinct combination process—it can be thought of as labelled with a parameter indicating the rate at which the process is taking place. Given the generation tree and an initial population of programs it is possible to determine the population of programs to be expected at any later time. A change in the rate of any combination process produces a change in the generated population sequence. That is, a change of any rate constitutes a change of generation procedure.

With this refinement, adaptation becomes a matter of modifying the parameters associated with the auxiliary vertices of the generation tree—the object being to change the generation procedure in such a way as to produce combinations of programs better suited to the environment. Assume that the set of generation procedures accessible to the adaptive system has been specified. Then the process of adaptation induces an orbit (in the set-theoretic sense) upon the set. Different systems of adaptation will induce different orbits. If the different orbits can be assigned ratings then this abstract basis can be used to formulate questions of adaptive efficiency.

The problem of assigning ratings to orbits will be discussed near the end of this section; there is, however, a prior problem: realization of generation procedures. As in information theory, realizations are of considerable importance—the channel capacity theorems gain much of their significance from the companion "realizability" theorems which state the existence of codes permitting transmission rates arbitrarily close to capacity (i.e., efficiencies arbitrarily close to 1). Models of the foregoing generation procedures can be constructed along the following lines:

The generators can be thought of as embedded in a discrete or cellular space, each type occurring with a given density (the expected number of generators in some fixed number of cells).

Each generator undergoes a random walk in the space. Upon coming into contact with another generator it may, with a probability determined by generators involved, connect to it (connected sets of generators correspond to programs). At the same time combinations of generators may, with a probability again determined by the types of generators involved, separate into component combinations. The rate at which generators come into contact (the generation rate) together with the connection and disconnection probabilities determine the density of each type of program as a function of time and the initial densities. That is, these factors determine what programs are generated and in what order. Each choice of a set of connection and disconnection probabilities selects a particular generation procedure. Special programs—so-called templates—can be added to the model to modify the probabilities, a given population of templates thus determining a unique generation procedure. Let the generation procedure associated with a given template-free model be called a *free generation procedure,* and let the result of adding some templates to the given model be called a *modified generation procedure.* If the free generation procedure is taken as a basis, then each modified generation procedure produces a population of programs skewed relative to that of the free procedure. If it is assumed that there is a "cost" involved in producing templates, then we can associate a "skewing cost" (per unit time) with each modified generation procedure.

The iterative circuit computers mentioned earlier, will be used to provide formal realizations of these cellular spaces, their laws and processes. The iterative circuit computer is not in itself an adaptive system; it should be identified, rather, with the space in which the adaptive system is embedded. The generators and their combinations will be represented by programs in the computer. Because the iterative circuit computers have been characterized mathematically, they can be used as a formal basis for theorems about the embedded systems.

Each computer in the class is constructed of a single basic module (a fixed logical network) iterated to form a regular array of modules. An appropriate choice of the basic module will yield an iterative circuit computer structurally and behaviorally equivalent to any of the following types of automata: Turing machines (with 1 or more tapes) [9] [8], tesselation automata [10] [7], growing logical nets [1] [2], and potentially-infinite automata [3]. Programs of an interative circuit computer can be given the following properties relevant to their interpretation as generators:

(1) A given iterative circuit computer can execute arbitrarily many sub-programs simultaneously (within limits imposed by size).

(2) Sub-programs can be written so that, under local control, they shift themselves from one set of modules to another set. Thus the geometry of the underlying iterative circuit computer becomes the geometry of the space in which the embedded program moves.

(3) Sub-programs which are independent initially can move into contact (occupy adjacent sets of modules) and connect so as to form a larger sub-program capable of moving and acting as a unit.

(4) Given any iterative circuit computer it is possible to select a finite set of sub-programs (individual instructions in the limiting case) to serve as generators such that any sub-program possible for that computer can be achieved by an appropriate combination of copies of these sub-programs.

(5) By a suitable choice of the basic module it can be arranged that generators do not interpenetrate (or "over-write") when shifting (cf. the notion of a "billiard ball" physics). Using the foregoing properties of generators one can realize any of the generation procedures described earlier in this section.

The question of rating orbits in the space of generation procedures can be treated now on the basis of the preceding discussion of realization. A given adaptive system can only be rated in terms of its performance in given environments. Thus some thought must be given first to characterization of permissible environments. It will only be noted here that an environment can be thought of as a population of problems presented to the system and that problems can be embedded in the same space as the generated programs. (That is, problems can be coded into the storage registers of the iterative circuit computer.) On their random walk through the space the generated programs will encounter and attempt to solve the embedded problems. In other words, the population of generated programs acts upon the population of problems to produce solutions.

Let each problem type be assigned a numerical quantity called "activation" (the quantity might also have been called "reward" or "utility of solution"). This quantity is to be consigned or "released" to the adaptive system whenever the associated generation procedure solves the problem by means of one of its programs. Let A(G, E, t) be the expected net rate of activation release at time t when the generation procedure G is faced with environment E. (The net rate can be taken as the expected release rate minus the skewing cost associated with G.) Two generation procedures confronted by the same environment over an interval of time, T, can then be compared in terms of the expected net activation release over that interval. One can assign ratings in a similar way to orbits in the space of generation procedures.

Briefly then the problem of adaptation, in the framework outlined, becomes one of modifying a free generation procedure in order to maximize activation release from the environment. In the paper referred to earlier [5] the rate at which a given adaptive system accumulates activation determines its survival (through duplication) in a population of adaptive systems. As a result the population of adaptive systems undergoes a differential selection according to ability to solve problems in the given environment. That system is fittest which in the long run accumulates the most activation.

4. Adaptive efficiency

4.1 Conditions, definitions, and initial considerations

The process of adaptation, as interpreted in section 3, involves a generated population of problem-solving programs acting upon a population of problems in an attempt to produce solutions. Assume that a set of generators, complete with respect to some universal computer, has been chosen and that the set of all possible programs formed from these generators has been enumerated and labelled $d_1, d_2, d_3, \ldots, d_i, \ldots$. Assume, furthermore, that a population of problems has been given and that it includes a denumerable number of different problem types $e_1, e_2, \ldots, e_j, \ldots$.

Following the description of section 3, let the programs and problems undergo random walks in the embedding space. We will concentrate our attention upon a bounded region of the space. Let the expected number of programs of type d_i in a unit volume at time t be a defined quantity, uniform over the bounded region (i.e., the programs are to be thought of as distributed with "uniform density"). Denote this quantity by $\rho(d_i)$; it will usually be referred to as the "density of d_i." Problems will be treated similarly and $\rho(e_j)$ will denote the "density of problems of type e_j." Let $\rho_d = \sum_i \rho(d_i)$, the "total program density," and $\rho_e = \sum_j \rho(e_j)$, the "total problem density," both be finite quantities. Let

$$\psi = \left[\frac{\rho(d_1)}{\rho_d}, \frac{\rho(d_2)}{\rho_d}, \ldots, \frac{\rho(d_i)}{\rho_d}, \ldots\right]$$

and

$$\theta = \left[\frac{\rho(e_1)}{\rho_e}, \frac{\rho(e_2)}{\rho_e}, \ldots, \frac{\rho(e_j)}{\rho_e}, \ldots\right].$$

ψ and θ, as defined, are elements of Hilbert space. In what follows the problem population θ is held constant while the program population ψ varies during the course of its adaptation to θ (ψ thus being a function of time).

For the purposes of this example, let each problem be so designed that only one program at a time can attempt its solution. A problem will be called "unoccupied" during any interval when no program is attempting its solution; at all other times it will be designated as "occupied." An individual program, during its random walk, will encounter various ones of the embedded problems. It will be assumed that each time the program encounters an unoccupied problem it attempts a solution. Such an encounter, between a program of type d_i and an *unoccupied* problem of type e_j, will be called an attempt of type (i, j). Each attempt of type (i, j) is to last for an expected time t_{ij}, at the end of which the expected activation release is a_{ij}. That is, the program d_i "attempts to solve" the problem e_j for t_{ij} units of time (on the average) and is "rewarded" by an amount a_{ij} (on the average) for its efforts. If d_i does not even partially solve e_j then $a_{ij} = 0$. The a_{ij} become progressively greater for program types which provide more and more extensive partial solutions, reaching a maximum for programs (if any) which solve the problem completely.

Because each attempt at a solution involves an "occupation time" t_{ij}, a law of diminishing returns applies to attempts. As $\rho(d_i)$ is increased, for any i, an increasing proportion of problems of any type e_j can be expected to be occupied by the d_i. Thus as $\rho(d_i)$ increases, an individual of type d_i will find fewer and fewer e_j unoccupied and, for that individual, the expected number of attempts of type (i, j) per unit time will decrease. But then, since activation release occurs only after an attempt, the expected rate of activation release *per individual* decreases also. In most cases of interest, the skewing cost per individual of a given type ultimately increases as one attempts to skew the generation procedure more and more to the generation of that one type. Under such conditions the net rate of activation release *per unit volume* for individuals of a given type will become *negative* as $\rho(d_i)$ approaches ρ_d. These conditions can be subsumed under the following general statement: A situation will be called *competitive* if, for any finite set of solvers D, the net rate of activation release per solver becomes negative as $\rho(D)$ approaches ρ_d.

It is important to note that, under competitive conditions, the discovery of *each* new $a_{ij} > 0$ is important to the adaptive system. For instance assume that at time t the system knows $a_{i_1 j_1} > 0$ and has just discovered $a_{i_2 j_2} = a_{i_1 j_1}$; assume further that d_{i_1} cannot solve e_{j_2} and rejects it immediately so that $a_{i_1 j_2} = 0$ and $t_{i_2 j_2} = 0$ and, similarly, $a_{i_2 j_1} = 0$, $t_{i_2 j_1} = 0$. Let $\rho(d_{i_1}) = \rho_0$ just before the discovery of $a_{i_2 j_2}$. Let $\rho(d_{i_1}) = \rho(d_{i_2})$

$= \rho_0/2$ after the discovery. Then other things being equal, a given program of type d_{i_1} (or d_{i_2}) will encounter an *unoccupied* problem of type e_{j_1} (or e_{j_2}) more frequently after the discovery than before. The increased attempt rate, coupled with the same total density of programs d_{i_1} and d_{i_2} and the fact that $a_{i_1 j_1} = a_{i_2 j_2}$, yields an increased expected activation release. (A similar advantage accrues when it is found that a single program can solve more than one problem.) Briefly, each new source, $a_{ij} > 0$, allows reduction of the saturation effect by allowing the successful programs to be "spread" over a larger number of sources.

In the idealized situation so far presented, complete information about the environment would consist of complete knowledge of the matrix a_{ij}, the matrix t_{ij}, and the population θ. The adaptive system possessing this information could then skew its program population ψ to maximize net activation release. Lacking such information, the system can acquire it only through encounters between elements of its program population ψ and the problem population θ. Obviously the system can only use information already accumulated as a basis for skewing ψ. For example, consider a bounded adaptive system which initially has no information about its environment. As time passes encounters between its generated programs and the problems will provide information about some of the a_{ij} (and the associated t_{ij}). After a time t, if this information is accumulated, some finite subset, D(t), of program-types will be *known* to effect a positive release of activation from one or more problem types. The system can then modify its generation procedure accordingly.

One strategy the adaptive system can employ in adapting to the environment is to adjust ψ at time t so that *net* activation release from *known* values of a_{ij} is a maximum. That is, the density of elements of $\psi(t)$ can be skewed so that activation release expected from programs belonging to D(t), *minus* the cost of skewing ψ, is a maximum. Under competitive conditions ρ_D, the total density of programs belonging to D(t), will be less than ρ_d.

We can show that this strategy is a "minimax" procedure for handling the environment. To see this, let a loss function $\lambda(\psi, \theta)$ be defined in the following terms:

$$A(\psi,\theta) \overset{\text{df.}}{=} \text{the expected net activation release (per unit time-volume) when program population } \psi \text{ operates on problems population } \theta$$

$$\mu(\theta) \overset{\text{df.}}{=} \psi_\theta \text{ such that } A(\psi_\theta, \theta) = \max_{\psi} A(\psi,\theta)$$

$$\lambda(\psi,\theta) \overset{\text{df.}}{=} A(\mu(\theta),\theta) - A(\psi,\theta)$$

In other words, the loss assigned to ψ in the presence of θ is the difference between the maximum net activation release possible and that actually obtained by ψ.

$\{\theta\}_t \overset{\text{df.}}{=}$ the set of problem populations having a_{ij} compatible with the values of a_{ij} *known* at time t.

$\psi_t \overset{\text{df.}}{=}$ the skewed program population resulting from the strategy of maximizing net activation release from the *known* a_{ij}.

$A_D(\psi,\theta) \overset{\text{df.}}{=}$ the expected net activation release from program types belonging to the subset D when ψ operates on θ (where D is some subset of the set of all program types, $\{d_i\}$)

Consider some $\psi \neq \psi_t$. By definition of ψ_t

$$A_{D(t)}(\psi,\theta) \leqq A_{D(t)}(\psi_t,\theta)$$

For any ψ, D(t), and $\epsilon > 0$ it is always possible to chose $\theta' \in \{\theta\}_t$ such that

$$A(\psi,\theta') - A_{D(t)}(\psi,\theta') < \epsilon$$

(Consider those problems $e_{j'}$ such that, for all i either $a_{ij'} = 0$ or else $a_{ij'}$ is unknown; choose $\theta' \in \{\theta\}_t$ such that the $e_j{}'$ are in fact solvable only by programs which occur with very low density in ψ.) But then

$$\min_{\{\theta\}_t} A(\psi,\theta) \leqq \min_{\{\theta\}_t} A(\psi_t,\theta) \text{ for any } \psi$$

(Actually "min" should be replaced by "glb" in some cases.) Therefore

$$\max_{\{\theta\}_t} [A(\mu(\theta),\theta) - A(\psi,\theta)] \geqq \max_{\{\theta\}_t} [A(\mu(\theta),\theta) - A(\psi_t,\theta)]$$

Or
$$\max_{\{\theta\}_t} \lambda(\psi,\theta) \geqq \max_{\{\theta\}_t} \lambda(\psi_t,\theta) \text{ for any } \psi$$

Or
$$\min_{\psi} \max_{\{\theta\}_t} \lambda(\psi,\theta) = \lambda(\psi_t,\theta)$$

Thus ψ_t "minimaxes" the loss function.

Recall that ψ_t is actually the program population which results if the adaptive system takes a very short term or opportunistic view of the environment—attempting to maximize net activation release from known sources without taking into account unknown sources. On the other hand, a minimax procedure is by its very nature a very conservative procedure based upon the most pessimistic estimates of the future. That the same population, ψ_t, results whether the adaptive system operates on a very opportunistic basis or upon a very conservative basis was at least unexpected. Nevertheless the adaptive system can do better than ψ_t.

4.2 Rate of information acquisition vs. limiting rate of adaptation, for homogeneous systems

The fact that ψ_t results from a minimax strategy in itself suggests that there are more efficient adaptive procedures--generally, a performance much better than minimax can be achieved at some small risk (often the strategy can be designed to make the risk as small as desired while obtaining significant improvement). As mentioned in section 3, two adaptive systems in a given environment are to be compared (over some interval of time T) in terms of their expected accumulations of activation. Thus our search is for a strategy which, over an interval of time T, will accumulate more activation than the strategy yielding ψ_t.

The possibility of improvement stems from the adaptive system's ability to control the rate at which it receives information from the environment. During any interval, say from t to $t+\Delta t$, the system can be expected to attempt some program-problem combinations (d_i, e_j) for the first time. These first attempts permit the system to add to its list of known a_{ij}. The system can control the expected number of first attempts in a given interval by controlling ψ. If ψ is changed so that the density of program d_i is increased, then the expected number of encounters between d_i and various problems, e_j, will be increased and, as a result, the number of first attempts involving d_i will also increase. In effect, the first attempts constitute a random sampling of the environment biased according to ψ. Note that if both ψ and θ are fixed, the expected number of first attempts steadily decreases; to maintain a fixed (expected) rate of first attempts the adaptive system must continually modify ψ.

Let any encounter resulting in a first attempt be called an *informative encounter* and let all other encounters be called *non-informative*.

$c(t) \overset{\text{df.}}{=}$ the expected number of encounters in the interval $(t, t+\Delta t)$

$$\gamma(t) \overset{\text{df.}}{=} \text{the proportion of encounters in the interval } (t, t+\Delta t) \text{ expected to be informative encounters.}$$

By suitably controlling the population ψ the adaptive system can control γ and hence the rate at which it acquires new information about the environment. In what follows it will be assumed that the average informative encounter itself contributes negligibly to the expected activation release—negligibly because of the inefficiency of the exploration process in comparison with the process aimed at releasing activation from known sources. However the population ψ, while maintaining γ, can be modified to take advantage of the new information. That is, at any given time there are many populations ψ which will yield the proportion γ; certain of these populations use the newly discovered a_{ij} to better advantage than the population based upon the old list of known a_{ij}. By so modifying ψ in terms of the new information, the expected activation release, averaged over all non-informative encounters, will be incremented.

$$\alpha(t) \overset{\text{df.}}{=} \text{the expected activation release per non-informative encounter at time t.}$$

$$\delta(t) \overset{\text{df.}}{=} \text{the } \textit{average} \text{ increment in } \alpha(t) \text{ to be expected from use of information from a single informative encounter at time t.}$$

In terms of the above functions:

$$\alpha(t) = \alpha(0) + \int_0^t \delta\gamma c\,dt$$

If $\eta(t)$ is the skewing cost at time t, then the net activation release (per unit time-volume) at time t, A(t), is given by:

$$A(t) = \alpha(1-\gamma)c-\eta$$

$$\text{Let } A^*(T) = \int_0^T A(t)\,dt$$

$$= \int_0^T [\alpha(1-\gamma)c-\eta]\,dt$$

In general, of two homogeneous bounded adaptive systems of the same size, the one producing the greater $A^*(T)$ will have a selective advantage. The function γ is the control function which the system adjusts in an attempt to optimize $A^*(T)$ for the environment θ. To gain some idea of how γ affects $A^*(T)$ let us impose the following simplifications:

Let θ and hence c be constant. Constrain the adaptive system to operate with γ set to a constant $\bar{\gamma}$. Assume an average $\delta(t) = \bar{\delta}$ can be defined for $\delta(t)$ when $\gamma(t) = \bar{\gamma}$ and let $\eta(t) = k\bar{\gamma}c$ (the greater the number of informative encounters per unit time, the greater the skewing cost). Under these conditions

$$\alpha(t) = \alpha(0) + \int_0^t \overline{\delta\gamma}c\,dt = \alpha(0) + \overline{\delta\gamma}ct$$

and

$$A^*(T) = \int_0^T [(\alpha(0) + \overline{\delta\gamma}ct)(1-\bar{\gamma})c - k\bar{\gamma}c]\,dt$$

$$= \alpha(0)(1-\bar{\gamma})cT + \overline{\delta\gamma}(1-\bar{\gamma})c^2\frac{T^2}{2} - k\bar{\gamma}cT.$$

The constant $\bar{\gamma}$ which maximizes $A^*(T)$ can now be determined; for a maximum

$$\frac{dA^*}{d\bar{\gamma}} = 0$$

or

$$-\alpha(0)cT + \bar{\delta}c^2\frac{T^2}{2} - \overline{\delta\gamma}c^2T^2 - kcT = 0$$

whence

$$\bar{\gamma}_0 = \left[\left(\frac{1}{2} - \frac{k+\alpha(0)}{\bar{\delta}cT}\right), 0\right]_{max}$$

For many interesting systems, T is determined by the time the system requires to accumulate some fixed amount of activation A_0^*. Upon accumulating A_0^* the system duplicates and both it and its offspring begin anew to accumulate A_0^*. If each of two systems requires A_0^* to duplicate, then the one with the smaller T will have a selective advantage. If, in addition, each system has the same small probability $p>0$, over an interval Δt, of disintegrating into its component parts, then the one with the smaller T will eventually displace the other completely in a mixed population of the two systems. γ_0, of course, yields the smallest possible T under the constraints imposed upon its calculation.

If $\bar{\delta}cT$ (the increment in α to be expected over time T if all encounters were informative) is large compared to $k+\alpha(0)$ (the skewing cost per informative encounter added to activation released initially per non-informative encounter) then $\gamma_0 \cong 1/2$. Thus, under such conditions, the adaptive system's effort should be divided equally between obtaining new information and making use of information already available. If such a result were to hold under more general conditions, say under quasi-economic conditions, it would certainly be provocative—no economic entity spends anywhere near this much of its effort in research.

Attention should be drawn to the fact that this result was obtained under the assumption that the population ψ could be instantaneously modified to whatever form, under the constraint $\bar{\gamma}$, best suited the current set of known a_{ij}. Thus the result is a limiting result. If the time to adjust ψ is not negligible the performance will be degraded accordingly.

It should also be noted that the result was obtained under the assumption that the bounded adaptive system was homogeneous. If non-homogeneous systems are employed, improvements can be effected. In particular one can employ boundaries (connected sets of generators) which selectively pass certain problem types. In this way the local density of given problem types can be increased within the bounded region. This makes possible an increase in the efficiency (attempt rate) of any program which solves one or more of the concentrated problems. (A more detailed discussion of non-homogeneous systems, including some results indicating the conditions under which improvement can be expected, will be published in another paper. One such result indicates the desirability of a high surface-to-volume ratio for the bounded regions.)

5. Summary

At first sight one might assume that an adaptive system can do no better than to use discovered sources of positive utility (activation, reward) as fully as possible from the moment of discovery onward. Such a strategy is a "minimax" strategy for the restricted class of systems investigated in this paper and, as it turns out, it is not in general the best strategy available to such systems. A fortiori, it fails as a "universal" strategy for adaptation.

The crucial step in determining a more efficient strategy comes when one tries to ascertain the net value, to the system, of additional information. For the systems considered, the gross value of additional information is set by the *increment in activation release per operation* which can be expected from the use of the information. (Such a valuation can be used for any system which by its operations can accrue positive utility from its environment.) In a competitive situation (where the availability of unoccupied sources is a function of program density) each new source discovered can contribute to the increment in activation release per operation. Thus, under competitive conditions, information about a new source, or information about a single program which handles several known sources, or information about a program which handles a known source more rapidly, all has a non-zero gross value. To determine the *net* value of the information its cost must also be taken into account. The cost arises because the system must make less than full use of known sources in order to obtain additional information; the loss thus incurred is a cost which must be made up via use of the additional information. For the systems studied this cost arises when the generated program population is diverted from utilization of known sources (non-informative contacts) so as to increase the frequency of *new* program-problem combinations (informative contacts). The extent of this diversion is measured by the ratio, γ, of informative contacts to total contacts. The cost of maintaining γ at a given level is not a one-shot cost—if the generated population of programs remains constant, the number of informative contacts per unit time falls off exponentially. Under the conditions stated, each setting of γ corresponds to an acceleration of the rate of accumulation of activation. Now, one adaptive system is more efficient than another, in a given environment, if it can be expected to accumulate some critical amount of activation in a shorter time than the other. In systems for which γ is defined, the efficiency of the system, and its advantage under differential selection, thus depends upon the setting of γ (optimum values of γ have been determined for an elementary class of bounded homogeneous systems). On a broader scale, when gross value and cost are figured on the basis given, the net value of information is directly related to the selective advantage that information (used optimally) will give to the system.

Arguments similar to those used in the body of the paper can also be applied in more general cases. For example, if the problems, e_j, are stochastic processes which the programs, d_i, are to predict, the argument goes through much as before except that the frequency of various program-problem combinations becomes important because the activation release associated with each combination is a random variable which must be estimated. The results also extend naturally to certain non-homogeneous systems, systems which are partitioned into subsystems by means of selective boundaries permitting passage of some programs and problems but not others. Such considerations lead directly to the study of coupled generation procedures and the study of the capabilities of systems realizing such procedures. It can be shown that non-homogeneous systems will in general have a selective advantage over homogeneous systems,

The results presented are, in themselves, quite elementary, although they may be suggestive to those inclined to view them as signposts; the various relations and definitions, though exhibited in this elementary context, have a broader scope. The primary purpose of the paper, however, has been to give evidence that it is possible to obtain results pertinent to adaptation within a framework provided by automata theory, even when only elementary methods are employed. It does not seem unreasonable that more sophisticated methods applied along similar lines will extend the results both in breadth and depth.

Acknowledgement

I would like to thank the members of the Logic of Computers Group at The University of Michigan for several helpful discussions of parts of this paper. The work it represents has been supported by the United States Army Signal Corps through contracts DA-36-039-sc-87174 and DA-36-039-sc-89085.

Bibliography

1. Burks, A. W., "Computation, Behavior and Structure in Fixed and Growing Automata," *Self-Organizing Systems,* Pergamon Press (1960).
2. Burks, A. W., and Hao Wang, "The Logic of Automata," *J. Assoc. Computing Machinery, 4,* 2 and 3, 193-218, 279-297 (1957).
3. Church, A., "Application of Recursive Arithmetic to the Problem of Circuit Synthesis," *Summer Inst. Symb. Logic Summaries, 1957,* Institute for Defense Analysis (1957).
4. Fisher, R. A., *The Genetical Theory of Natural Selection,* Dover, (1958).
5. Holland, J. H., "Outline for a Logical Theory of Adaptive Systems," *J. Assoc. Computing Machinery, 9,* 3 (1962).
6. Holland, J. H., "Iterative Circuit Computers," *Proc. 1960 Western Joint Computer Conf.,* 259-265 (1960).
7. Moore, E. F., "Machine Models of Self-Reproduction," *Proc. Sympos. Math. Prob. Biol. Sci.,* New York (1961).
8. Rabin, M. O., and D. Scott, "Finite Automata and Their Decision Problems," *IBM J. Res. and Dev. 3,* 3, 210-229 (1959).
9. Turing, A. M., "On Computable Numbers, with an Application to the Entscheidungsproblem," *Proc. London Math. Soc.* (2), *43,* 230-265 (1936).
10. von Neumann, J., *The Theory of Automata: Construction, Reproduction, Homogeneity.* (unpublished manuscript).

UNIVERSAL SPACES: A BASIS FOR STUDIES OF ADAPTATION

John H. Holland

The University of Michigan
July 1964

The first objective of these lectures is to present a definition of automata appropriate to studies of construction and adaptation. This entails the development, in the context of automata theory, of three general ideas:

(1) structure, particularly that part of the concept relevant to connected sets of components having an organization presentable in terms of a hierarchy of block diagrams;

(2) computation, especially the computation procedures which can be realized by finite automata;

(3) simulation, the representation of details of the behavior of one device by appropriately constraining (programming) the behavior of another.

In what follows the formal counterparts of these ideas will be designated 'compositions', 'uniform computation procedures', and 'embeddings' respectively. Only a resume of these ideas is presented here; a full development appears in [6].

In terms of these definitions we can carry out the second step of the development: definition and investigation of the class of universal embedding spaces. Each such space will be defined in terms of a countably infinite, connected set of automata and will have the following properties:

(1) ability to represent both structure and behavior,

(2) homogeneity in the sense that, if an automaton has a representation in the space, it has the same representation everywhere in the space,

(3) universality in the sense that all computation procedures can be represented in the space (by embedding an appropriate composition).

Again only a resume is given, see [6] for a more complete development.

The last of the lectures will give a formal outline of the use of these notions in the study of adaptive systems.

—— * ——

I will begin with the notion of a composition. The concept is developed first by defining a "free" product automaton, which is simply a collection of non-interacting automata treated as a single automaton. Constraints are then applied to the product automaton by identifying (connecting) selected inputs with outputs. Formally, the latter process amounts to adding a set of constraining equations to the set of equations which defines the product automaton. The major difficulty will be that of assuring the consistency of the resulting set of simultaneous equations. The set of compositions will be the set of all consistently constrained product automata over finite or countably infinite sets of automata. Thus, corresponding to each finite automaton there will be an infinite class of compositions (structural representatives). Compositions over a countably infinite set of automata are an extension of the notion of finite automaton and it is from them that we shall draw the universal embedding spaces.

A word of motivation: although it would be impossible to present (by a state diagram, say) or deal directly with a transition function defined over an unstructured set of 1,000,000 states, that same transition function may easily be comprehended in terms of a connected set of 20 2-state devices. By the same token, though a state reduction from 1,000,000 states to 500,000 states looks impressive, it amounts to the elimination of only one 2-state device from the connected set. In some sense the only way of understanding complex devices (even a small digital computer will have in excess of 10^{10^4} states) is to present them in terms of a hierarchy of block diagrams: a diagram at level k being expanded into, say, 10 diagrams at level k - 1; the diagrams at level 0 consisting then of the basic components. In this way one can quickly investigate the behavior of any part of a device with 10^{10} basic components (for 2-state devices: $2^{10^{10}}$ states) using only 10 levels in the diagram hierarchy. In what follows the hierarchy of block diagrams will have as their formal counterpart a hierarchy of sub-compositions: the quintuple characterizing a sub-composition, at any given level, being structured in terms of the quintuples of sub-compositions at the next lower level.

Compositions will be defined for any indexed set of elements $\{a_\alpha \ni \alpha \varepsilon A\}$ satisfying the following conditions:

A is a countable ordered set

$a_\alpha = < I_\alpha, S_\alpha, O_\alpha, f_\alpha, u_\alpha >$

$I_\alpha = \prod_{i=1}^{m_\alpha} I_{\alpha,i}$ where each $I_{\alpha,i}$ is a finite set

S_α , a finite set

O_α , $\prod_{j=1}^{n} O_{\alpha,j}$ where each $O_{\alpha,j}$ is a finite set

$f_\alpha : I_\alpha \times S_\alpha \longrightarrow S_\alpha$

$u_\alpha : I_\alpha \times S_\alpha \longrightarrow O_\alpha$

Each quintuple constitutes one of the standard definitions of a finite automaton with f_α and u_α being interpreted as the transition and output functions respectively. f_α and u_α can be extended in the useful way to two families of functionals, F_α and U_α, on infinite input sequences:

$$F_\alpha : I_\alpha^N \times S_\alpha \longrightarrow S_\alpha^N$$
$$U_\alpha : I_\alpha^N \times S_\alpha \longrightarrow O_\alpha^N$$

where $N = \{0, 1, 2, \ldots\}$ and $I_\alpha^N = \{\underline{I}_\alpha \ni \underline{I}_\alpha : N \longrightarrow I_\alpha\}$, etc.

An unrestricted (product) composition, $\underline{A}$, on $\{a_\alpha\}$ is a quintuple $< I_{\underline{A}}, S_{\underline{A}}, O_{\underline{A}}, f_{\underline{A}}, u_{\underline{A}} >$ such that:

$$S_{\underline{A}} = \prod_{\alpha\varepsilon A} S_\alpha = \{s : A \longrightarrow \bigcup_\alpha S_\alpha \ni \alpha \varepsilon A \text{ and } s(\alpha) \varepsilon S_\alpha\} \quad ;$$

similarly $$I_{\underline{A}} = \prod_{\alpha\varepsilon A} \prod_{i=1}^{m_\alpha} I_{\alpha,i} = \{i : X \longrightarrow \bigcup_X I_X \ni x \varepsilon X \text{ and } i(x) \varepsilon I_X\}$$

where $X = \{(\alpha, i) \ni \alpha \in A \text{ and } 1 \leq i \leq m_\alpha\}$;

and $$O_{\underline{A}} = \prod_{\alpha \in A} \prod_{j=1}^{n_\alpha} O_{\alpha,j} = \{o : Y \longrightarrow \bigcup_y O_y \ni y \in Y \text{ and } o(y) \in O_y\}$$

where $Y = \{(\alpha, j) \ni \alpha \in A \text{ and } 1 \leq j \leq n\}$.

$f_{\underline{A}} : I_{\underline{A}} \times S_{\underline{A}} \longrightarrow S_{\underline{A}}$ satisfies the requirement

$$f_{\underline{A}}(i,s)(\alpha) = s'(\alpha) = f_\alpha(i(\alpha), s(\alpha))$$

where $i \in I_{\underline{A}}$, $s, s' \in S_{\underline{A}}$

and $i(\alpha) = (i(\alpha,1), \ldots, i(\alpha, m_\alpha))$.

$u_{\underline{A}} : I_{\underline{A}} \times S_{\underline{A}} \longrightarrow O_{\underline{A}}$ satisfies

$$u_{\underline{A}}(i,s)(\alpha) = o(\alpha) = u_\alpha(i(\alpha), s(\alpha))$$.

Each composition over $\{a_\alpha\}$ will be defined in terms of a composition function Υ satisfying the conditions:

(1) $\Upsilon : Y' \longrightarrow X'$ from $Y' \subset Y$ 1 to 1 onto $X' \subset X$

(2) $O_y \subset I_{\Upsilon(y)}$

Υ induces a set of equations of the form

$$U_y\ (I_\alpha,\ s) = I_{\Upsilon(y)} \qquad (*)$$

This set of constraining equations combined with the defining equations for the U_y gives us a set of simultaneous equations for which we require a unique solution. (That is, if such a solution exists we can go on to find a quintuple corresponding to the product automaton constrained by Υ.) To assure consistency of the equations (*), Υ must satisfy a further requirement of "local effectiveness":

Υ will be called consistent with respect to $s \in S_{\underline{A}}$ if for each $\alpha \in A$, there is an integer $\ell_{s,\alpha}$ such that every connected sequence of output indices (defined as usual, see [3]) to α of length $\ell_{s,\alpha}$ contains an element $y_h = (\delta, j)$ such that

$$(i_1, i_2 \varepsilon I_{\underline{A}})[i_1|X-X' = i_2|X-X' \Rightarrow u_{\delta,j}(i_1(\delta), s(\delta)) = u_{\delta,j}(i_2(\delta), s(\delta))]$$

That is, for state s, $u_{\delta,j}$ does not depend on any of its constrained inputs.

Define $I_{\Upsilon,\underline{A}} = \{i|X-X' \ni i \ \varepsilon \ I_{\underline{A}}\}$

and $O_{\Upsilon,\underline{A}} = \{o|X-X' \ni o \ \varepsilon \ O_{\underline{A}}\}$

calling these <u>composition</u> <u>input</u> (<u>output</u>) <u>states</u>.

Modifying an algorithm of Burks and Wright [3], one can prove the following:

<u>LEMMA</u>. If Υ is consistent with respect to s ε $S_{\underline{A}}$, then there is a unique extension $\mu_s : I_{\Upsilon,\underline{A}} \longrightarrow I_{\underline{A}}$ satisfying equations (*).

Υ will be called a locally effective composition function (LECF) if:

(1) Υ is consistent for some computable s ε $S_{\underline{A}}$,

(2) if Υ is consistent for s ε $S_{\underline{A}}$ then for every i ε $I_{\Upsilon,\underline{A}}$ Υ is consistent with respect to $f_{\underline{A}}(\mu_s(i), s)$.

For each LECF Υ and unrestricted composition $\underline{A}$ the <u>composition</u> $\underline{A}_{\Upsilon}$ is defined by the quintuple

$$< I_{\Upsilon,\underline{A}}, S_{\Upsilon,\underline{A}}, O_{\Upsilon,\underline{A}}, f_{\Upsilon,\underline{A}}, u_{\Upsilon,\underline{A}} >$$

where $I_{\Upsilon,\underline{A}}$ and $O_{\Upsilon,\underline{A}}$ are as above

$$S_{\Upsilon,\underline{A}} = \{s \ \varepsilon \ S_{\underline{A}} \ni \Upsilon \text{ is consistent w.r.t. s and s is computable}\}$$

$$f_{\Upsilon,\underline{A}}(i,s) = f_{\underline{A}}(\mu_s(i), s)$$

$$u_{\Upsilon,\underline{A}}(i,s) = u_{\underline{A}}(\mu_s(i), s) \text{ for } i \ \varepsilon \ I_{\Upsilon,\underline{A}}, \ s \ \varepsilon \ S_{\Upsilon,\underline{A}}$$

(The subscript $\underline{A}$ will be dropped where no confusion can arise.) Thus the set $\{\underline{A}_{\Upsilon} \ni \underline{A}$ is an unrestricted composition and Υ is LECF over $\underline{A}\}$ will be the set of all compositions.

<u>LEMMA.</u> Given any $\underline{A}_\gamma$, $\alpha \varepsilon A$, and $t \varepsilon N$: $\underline{S}_\gamma(t)|\alpha$ can be calculated from $\underline{S}_\gamma(o)|B(\alpha,t)$ and $\underline{I}_\gamma(o)|B(\alpha,t), \ldots, I_\gamma(t)|B(\alpha,t)$ where B is a computable function and $B(\alpha,t)$ is a finite subset of A containing α.

<u>THEOREM.</u> The property LECF is decidable for all Υ over finite A but not for all Υ over all A.

The following definitions are useful:

$\{a_1, \ldots, a_k\}$ will be called a set of <u>generators</u> for $\underline{A}_\gamma$ if $\{a_\alpha \ni \alpha \varepsilon A\}$ consists only of copies of elements of $\{a_1, \ldots, a_k\}$.

B_ζ over $\{b_\beta \ni \beta \varepsilon B\}$ will be called a <u>subcomposition</u> of $\underline{A}_\gamma$ over $\{a_\alpha \ni \alpha \varepsilon A\}$ if:

(1) $\{b_\beta\} \subset \{a_\alpha\}$

(2) $\zeta = \Upsilon|Y_B$ where $Y_B = \{y \varepsilon Y' \ni \text{proj}_1\ y \varepsilon B \text{ and } \text{proj}_1\ \Upsilon(y) \varepsilon B\}$.

For any composition $\underline{A}_\Upsilon$ one can construct a hierarchy of subcompositions corresponding exactly to any set of block diagrams describing the composition.

The idea of computation, for compositions, will be developed for functions $\Gamma : \Sigma_i \longrightarrow \Sigma_j$ where $\Sigma_i = \{\sigma : N \longrightarrow N_i \ni N = \{0, 1, 2, \ldots\}$ and $N_i = \{0, 1, 2, \ldots, i\}\}$. $\underline{A}_\Upsilon$ with initial state s will be said to compute Γ b-uniformly if: After any arbitrary 'start-up' time τ, and a fixed 'processing delay' c, $\underline{A}_\Upsilon$ produces successive values of $\Gamma(\sigma)$ every b units of time when the input sequence $\underline{I}_\Upsilon$ presents successive values of σ at the same rate. More formally:

$$(\exists c \varepsilon N)(\exists \Omega \varepsilon I_\gamma)(\forall \ \varepsilon \Sigma_i)(\forall n \varepsilon N)(\forall \tau \varepsilon N)$$

$$\Big[U_\Upsilon(\underline{I}_\Upsilon, s);\ (t + c) = \begin{cases} \Gamma\sigma(n) \text{ when } t = bn + \tau \\ \Omega \text{ otherwise} \end{cases}$$

$$\text{when } \underline{I}_\Upsilon(t) = \begin{cases} \sigma(n) \text{ for } t = bn + \tau \\ \Omega \text{ otherwise} \end{cases} \Big] .$$

Ω plays the role here of a 'no-signal' condition. (Actually this definition should be weakened to allow the input and output sets, I_γ and O_γ, to contain coded representations of elements of N_i and N_j rather than only the elements themselves).

In what follows it will be assumed that, whatever notion of computation is employed, the set of computable Γ will contain the set of b-uniformly computable Γ.

It is worth noting that, interpreting the sequences σ as exapnsions of real numbers, the uniformly computable Γ are continuous although often undefined at various points — several interesting questions arise, but they will not be followed up here.

<u>LEMMA</u>. If Γ is finitely computable in the sense of Burks [2] then Γ is uniformly computable.

<u>LEMMA</u>. If $\underline{A}_\gamma$ and $\underline{A}'_{\gamma'}$ b-uniformly compute Γ and Γ', respectively, and if $\Gamma\Gamma'$ (the composition of the functions) is defined, then $\underline{A}_\gamma$ and $\underline{A}'_{\gamma'}$ can be connected to form a new composition such that

$$U_\gamma(U_{\gamma'}(\underline{I}_{\gamma'}\, s),\ s') = \Gamma(\Gamma'(\sigma)) \text{ when } \underline{I}_\gamma(t) = \begin{cases} \sigma(n) & \text{for } t = bn + \tau \\ \Omega & \text{otherwise} \end{cases}.$$

Third in the line-up of underlying ideas is that of simulation. To specify the manner in which one device simulates another is to supply a mapping whereby actions in the image device can be reinterpreted as actions of the object being simulated. The formal counterpart of this will be an embedding map defined as follows:

Let A, X, Y and S_γ, I_γ, O_γ be the index and state sets, respectively, of the object composition, $\underline{A}_\gamma$, and let B, X_1, Y_1 and S_{γ_1}, I_{γ_1}, O_{γ_1} be the corresponding sets for the image composition, B_{γ_1}. A mapping φ from the sets

$A, X, Y, S_\gamma, I_\gamma, O_\gamma$ to subsets of $B, X_1, Y_1, S_{\gamma_1}, I_{\gamma_1}, O_{\gamma_1}$, respectively, will be called a <u>strict</u> <u>embedding</u> if:

(1) Distinct {indices, states} map onto distinct {sets of indices, states}:

$$\alpha \neq \alpha_1 \Longrightarrow \varphi(\alpha) \cap \varphi(\alpha_1) = \text{null set, etc.}$$

(2) Each index of a {free, bound} {input, output} of a given element maps onto corresponding indices of the image subset:

$$\text{proj}_1\, \varphi(x) \subset (\text{proj}_1\, x),\ x \in X - X' \Longrightarrow \varphi(x) \subset X_1 - X_1'$$

and similarly for y.

(3) If $x = \gamma(y)$ then the same must hold for all indices in $\varphi(x)$:

$$\varphi(\gamma(y)) = \gamma_1(\varphi(y)) \quad .$$

(4) If two states of $\underline{A}_\gamma$ assign the same state to an element, then the images of these states must assign the same state to the images of the element:

$$s(\alpha) = s_1(\alpha) \Longrightarrow \varphi(s)|\varphi(\alpha) = \varphi(s_1)|\varphi(\alpha)$$

and similarly for $i, i_1 \in I_\gamma$, $o, o_1 \in O_\gamma$.

(5) When the state of $\underline{B}_{\gamma_1}$ is the image, $\varphi(s)$, of a state s of $\underline{A}_\gamma$, then f_{γ_1} does not depend upon the states of unassigned inputs:

$$i|\varphi(X - X') = i_1|\varphi(X - X') \Longrightarrow f_{\gamma_1}(i, \varphi(s)) = f_{\gamma_1}(i_1, \varphi(s))$$

where $i, i_1 \in I_{\gamma_1}$.

(6) The successor states of the element images must be the images of the successor states:

$$f_{\varphi(\alpha)}(\varphi(i)|\varphi(\alpha), \varphi(s)|\varphi(\alpha)) = \varphi f_\alpha(i|\alpha, s|\alpha)$$

where $f_{\varphi(\alpha)}$ is the transition function of the sub-composition indexed by $\varphi(\alpha)$.

Conditions (5) and (6) assure that the transformations, $F_{\varphi(\alpha)}$ and $U_{\varphi(\alpha)}$, of the image sub-compositions can be restricted so as to be the same as F_α and U_α .

The requirements on φ can be weakened in a natural way to yield weak and b-slow embeddings or, briefly, embeddings:

[weak] (1) $\underline{A}$ can be partitioned into disjoint sub-compositions which are then embedded as elements, only overall behavior of the sub-compositions being preserved (thus, for example, Υ need not be preserved in the sub-compositions).

[slow] (2) A b-slow embedding yields an image with a time scale t' = bt where t is the time index of $\underline{A}_\Upsilon$. This can be accomplished in a natural way by mapping input states of the object composition into strings of input states in the image. That is, input sequences to the object composition are "slowed" in the image by the "insertion" of b - 1 'no-signal' states between signals, cf. McNaughton [9].

The characterization of homogeneity for compositions, which I will discuss next in conjunction with universal compositions, depends upon three definitions:

An isomorphic embedding, φ, of $\underline{A}_\Upsilon$ on $\underline{B}_{\Upsilon_1}$ is a strict embedding such that $\alpha \simeq \varphi(\alpha)$, $x \simeq \varphi(x)$, $y \simeq \varphi(y)$, $S_\alpha \simeq S_{\varphi(\alpha)}$, $I_x \simeq I_{\varphi(x)}$ and $O_y \simeq O_{\varphi(y)}$.

Two embeddings, φ_1, and φ_2, of $\underline{A}$ in $\underline{C}_{\Upsilon_2}$ {on a single generator} will be called identical {identically oriented} if:

(1) There is an isomorphic embedding, θ, of the sub-composition indexed by $\varphi_1(A)$ on the sub-composition indexed by $\varphi_2(A)$ such that $\theta\varphi_1 = \varphi_2$

(2) For all α, $\alpha' \in A$ and any connected sequence ρ from $\varphi_1(\alpha)$ to $\varphi_1(\alpha')$ {or from $\varphi_1(\alpha)$ to $\varphi_2(\alpha)$}, there exists a connected sequence ρ' from $\varphi_2(\alpha)$ to $\varphi_2(\alpha')$ {from $\varphi_1(\alpha')$ to $\varphi_2(\alpha')$} such that $\text{proj}_2\, \rho = \text{proj}_2\, \rho'$ (where $\text{proj}_2\, \rho = (i_1, i_2, \ldots, i_\ell)$ just in case $y_h = (\beta_h, i_h)$, $1 \leqslant h \leqslant \ell$)

——— * ———

A composition $\underline{V}_\nu$ will be called <u>universal and locally homogeneous</u> for (uniformly) computable Γ if:

(1) for each (uniformly) computable Γ there is an embedding φ, into $\underline{V}_\nu$, of a composition $\underline{A}_\gamma$ capable of (uniformly) computing Γ;

(2) if φ_1 embeds $\underline{A}_\gamma$ in $\underline{V}_\nu$ then, given arbitrary $\alpha \varepsilon A$, $\xi \varepsilon \varphi(\alpha)$ and $\xi' \varepsilon V$, there exists φ_2 embedding $\underline{A}_\gamma$ in $\underline{V}_\nu$ identically so that $\Theta(\xi) = \xi'$.

<u>LEMMA</u>. Necessary conditions for $\underline{V}_\nu$ to be universal and locally homogeneous are:

(1) $\underline{V}_\nu$ must be generated by a single element,

(2) V must be countably infinite.

If condition (2) of the definition is modified to require identical orientation (since in either case $\underline{V}_\nu$ must be generated by a single element) we obtain a third necessary condition:

(3) strings over output indices are "commutative", i.e. if ρ is a connected sequence from α to α', $\sigma = \text{proj}_2\,\rho$, and σ' is any permutation of σ, then there exists a connected sequence ρ' such that $\text{proj}_2\,\rho' = \sigma'$.

Compositions satisfying the modified definition will be called universal and homogeneous or, simply, <u>computation-universal</u>. It is a consequence of the third part of the lemma that computation-universal spaces can be "co-ordinatized" on a discrete (integer) cartesian k-dimensional grid ($k \leqslant n_\alpha$). In these terms the second part of the definition assures that all "translations" of an image are also images of the same object. As a consequence, properties of the image, such as its connection scheme, can be made independent of the "location" of the image.

Let U denote the class of computation-universal compositions. If part (1) of the definition of computation-universal is modified to read:

(1) given any finite composition $\underline{A}_\gamma$ there exists an embedding φ of $\underline{A}_\gamma$ in $\underline{V}_\nu$,

then a natural subclass, U_c, of U is defined — the class of composition-universal compositions. Another natural sub-class, U_d is defined by

$$U_d = \{\underline{V}_\nu \ni \underline{V}_\nu \text{ is computation-universal with a Moore (lag-time} \geqslant 1\text{) automaton as generator}\} .$$

Elements of U_d are the most natural generalizations of von Neumann's "logical-universal" space [11] because each exhibits the "propagation delay" which plays an important part in his study of self-reproduction. In fact von Neumann's detailed example establishes the existence of U_d and hence of U. It can also be shown that there exist elements in U_c (the class of iterative circuit computers [5] contains a subset of composition-universal compositions).

It can be established that U_d and U_c are disjoint — a result which has important consequences for the study of construction and adaptation via universal compositions:

THEOREM. Given any $\underline{V}_\nu \in U_d$, any set of automata, G, containing at least one automaton with two input dependent outputs, and $b \in N$, there exists a finite composition $\underline{A}_\gamma$ generated by G which cannot be b-slow embedded in $\underline{V}_\nu$. (In fact for each $\underline{V}_\nu$, G, and b there exists a constant c and a function $e(\ell)$ such that the theorem is true for any $\underline{A}_\gamma$ containing sub-composition "trees" with at least $e(\ell)$ elements, $\ell \geqslant c$; $e(\ell)$ and c are such that "almost all" $\underline{A}_\gamma$ over G satisfy the condition. The theorem turns on the failure of Kleene's representation theorem [8] for the embedded compositions.)

Thus the elements of U_d are not composition-universal; it follows at once that the generating element of any composition of type U_c must be a Mealy (lag-time zero for some states) automaton.

<u>COROLLARY.</u> Given any finite composition $\underline{A}_\gamma$ and any b-slow embedding of $\underline{A}_\gamma$ in a composition $\underline{V}_\nu \varepsilon U_d$, "almost all" compositions containing $\underline{A}_\gamma$ as a sub-composition cannot be b-slow embedded in $\underline{V}_\nu$.

Thus, given any finitely computable Γ, "almost none" of the compositions capable of computing Γ can be embedded in $\underline{V}_\nu \varepsilon U_d$. Moreover, given a set of computation procedures with a common "sub-routine" it will in general be impossible to embed them in $\underline{V}_\nu$ so as to preserve the common sub-routine. Or again:

<u>COROLLARY.</u> Given $\underline{V}_\nu \varepsilon U_d$, any $b \varepsilon N$, and any set of generators G, as above, there exist $\Gamma_1, \Gamma_2, \ldots, \Gamma_j, \ldots$ such that

(1) each Γ_j is b-uniform computable by some $[\underline{A}_j]\gamma_j$ over G;

(2) $[\underline{A}_j]\gamma_j$ can at best be b_j-slow embedded in V_ν,

where $b_j > b_{j-1}$, $b_j \varepsilon N$.

Since $\lim_j b_j = \infty$, the computation rates of the images approach zero.

I would say that the above results argue strongly for the use of spaces of type U_c, rather than U_d, in studies of construction and adaptation. Though one may question how "realistic" such spaces are in allowing propagation of signals with negligible delay, no mathematical difficulty ensues because the spaces U_c are compositions and hence satisfy strong local effectiveness conditions (see the lemma following the definition of composition).

—— * ——

I will turn now to adaptive systems. By way of motivation, note that there is a problem of adaptation only when some aspects of the environment are initially unknown to the adaptive system. If we think of the system as attempting to model the environment, then its model is incomplete. The system cannot predict the consequences of some of its responses; in terms of its model the system may be facing any of several environments determined by the possible alternatives (substitution instances) for unknown aspects. In what follows this will be formalized by designating a class, $\mathcal{E}$, of admissible (or possible) environments.

In a similar vein, it is necessary to specify the techniques or models available to the adaptive system before one has a clearly defined problem of adaptation. The class of devices which the system can bring to bear through rearrangement of its structure will be designated $\mathcal{A}$.

For present purposes both $\mathcal{E}$ and $\mathcal{A}$ will be taken to be possibly infinite collections of compositions (the elements of $\mathcal{A}$ being restricted to finite compositions). An adaptive strategy, then, can be thought of as a set of trajectories through the set $\mathcal{A}$, parametrized by the elements of $\mathcal{E}$. That is, the strategy will dictate successive reorganizations of the adaptive system (the trajectory) according to the information (inputs) it receives from the environment, $E \in \mathcal{E}$, confronting the system.

In greater detail:

$$\text{Let } \mathcal{E} = \{\underline{E}_b \ni \underline{E}_b = < I_b, S_b, f_b, s, q_b, \mu_b > \text{with } b \in B\} \quad ,$$

$$s \in S_b \text{ and } B \text{ an ordered index set.}$$

$$\text{Let } \mathcal{A} = \{\underline{A}_d \ni \underline{A}_d = < I_d, S_d, f_d, q_d > \text{with } d \in D\}$$

$$D \text{ an ordered index set.}$$

I_b, S_b, f_b and I_d, S_d, f_d will be defined as for compositions, q_b will be defined as a map

$$q_b : \bigcup_{d \in D} (I_d \times S_d) \longrightarrow I_b \quad .$$

Under interpretation, q_b determines the input (response) to environment $\underline{E}_b$ for any complete state of any admissible device in $\mathcal{A}$. Similarly

$$q_d : \bigcup_{b \varepsilon B} S_b \longrightarrow I_d \quad ;$$

q_d corresponds to the symbolization of the environment (cues, etc.) used by $\underline{A}_d$. It is worth noting that even if $\underline{A}_{d_1}$ differs from $\underline{A}_{d_2}$ only because $q_{d_1} \neq q_{d_2}$ (that is, as automata, $\underline{A}_{d_1}$ and $\underline{A}_{d_2}$ are essentially the same) the adaptability of $\underline{A}_{d_1}$ may be greatly different from that of $\underline{A}_{d_2}$. The amount of information $\underline{A}_d$ receives from the environment depends critically on q_d; changes in q_d may be an essential part of the system's adaptation to the environment.

Finally μ_b is a utility function over the states S_b of $\underline{E}_b$, $\mu : S_b \longrightarrow$ reals. In this formulation each state of the environment is assumed to yield some payoff (possibly zero) to the adaptive system (cf. the corresponding formulation in the theory of games, interpreting states as outcomes).

An adaptive strategy, τ, will be a map

$$\tau : D \times [\bigcup_{d \varepsilon D} I_d] \times [\bigcup_{d \varepsilon D} S_d] \longrightarrow D$$

which determines for every device and every complete state of that device what admissible device is to succeed it. That is, the device employed by the adaptive system at time t + 1 is determined from the complete state of the device employed at time t:

$$d(t+1) = \tau(d(t), i(t), s(t))$$

$$\text{where } i(t) \varepsilon I_{d(t)},\ s(t) \varepsilon S_{d(t)} \quad .$$

Given a device $\underline{A}_{d(o)}$ and its initial state, an environment, and an adaptive strategy τ, the overall behavior of the system is determined by the following recursions:

$$I(t) = q_{d(t)}(S_b(t)), \qquad I(t) \varepsilon I_{d(t)}$$

$$I_b(t) = q_b(I(t), s(t)), \qquad s(t) \varepsilon S_{d(t)}$$

$$S_b(t+1) = f_b(I_b(t), S_b(t)), \qquad S_b(o) = s$$

$$d(t+1) = \tau(d(t), I(t), S(t))$$

$$S'(t+1) = f_{d(t)}(I(t), S(t))$$

$S(t+1)$ is determined from $S'(t+1)$ by deleting those components of $S'(t+1)$ corresponding to components deleted by τ and adding components corresponding to the initial states of components added by τ. For example, if $\underline{A}_{d(t+1)}$ is a sub-composition of $\underline{A}_{d(t)}$, indexed by $A_1 \subset A$, then $S(t+1) = S'(t+1)|A_1$.

If the devices produced by strategy τ are embedded in a universal space $\underline{V}_\nu$ then a natural restriction on the "amount of construction per unit time" results:

Let $B(S(t), \alpha)$ be the set of all elements in $\underline{V}_\nu$ which can affect the state of element a_α at time $t + 1$, given the state assignment $S(t) \varepsilon S_\nu$ (cf. the definition: ν consistent with respect to $S(t)$).

Let $V(t)$ be the index set of elements in the sub-composition which is the image of $A_{d(t)}$ in $\underline{V}_\nu$ and let $\overline{V}(t) = \{\alpha \ni B(S(t), \alpha)$ contains $\beta \varepsilon V(t)\}$.

Require $V(t+1) \subseteq \overline{V}(t)$.

That is, τ at any given time can only add elements which, in the embedding, lie within the radius of immediate action (unit delay) of the extant image. When $\tau(d(t), I(t), S(t)) \neq d(t)$ this must be achieved by the image of $I(t)$, $\varphi(I(t))$, causing changes in the dependencies of the sub-composition indexed by $\overline{V}(t)$ so that $f_{V(t+1)}(\varphi(I(t)), \varphi(S(t))) = \varphi(f_{d(t+1)}(I(t), S(t)))$, whereas $f_{V(t)} = \varphi f_{d(t)}$ before.

From this point on it will be assumed that all strategies begin with some fixed device $\underline{A}_{d(o)}$ in state $S_{d(o)}(o)$. Under this assumption, τ when confronted by any environment E_b will determine a unique environmental sequence

$S_b(1)$, $S_b(2)$, ..., $S_b(t)$, ... This in turn will determine a sequence of pay-offs $\mu(S_b(1))$, $\mu(S_b(2))$, ... Let $\mu_{\tau,E_b}(t)$ designate the pay-off so determined by τ at t.

τ_o will be called ϵ-near-optimal if

$$(\forall E_b \varepsilon \mathcal{E})(\text{almost all admissible } \tau) \quad [\lim_{T\to\infty} \frac{\sum_o^T \mu_{\tau_o,E_b}(t)}{\sum_o^T \mu_{\tau,E_b}(t)} > \epsilon] \ .$$

Note that subset of $\{\tau\}$ for which the condition need not be met may vary from one element of $\mathcal{E}$ to another without destroying the near-optimality of τ_o. The motivation for this definition lies in its consequences: If such a τ_o exists for a pair $(\mathcal{E}, \mathcal{Q})$ then, given any $E_b \varepsilon \mathcal{E}$, almost no other strategy can force τ_o into gambler's ruin. (The exceptions are these enumerative strategies which happen to produce a device, early in the sequence $\underline{A}_{d(t)}$, which is optimal for E_b; if the strategy is enumerative, however, it will perform poorly over almost all other elements of $\mathcal{E}$, assuming $\mathcal{E}$ is at all large). Stated another way, if the payoff determines the duplication rate of an entity employing strategy τ, then an ϵ-near-optimal strategy will escape extinction against almost all other strategies in any environment of $\mathcal{E}$.

The definition may be strengthened to strictly-ϵ-near-optimal by substituting "$\forall\tau$" for "almost all τ". The difficulty of finding interesting pairs $(\mathcal{E}, \mathcal{Q})$ for which such strategies exist is accordingly increased.

The necessity for definitions of optimality, like the above, which involve "rate of accumulation" or "efficiency" is shown by the trivial way in which "convergence to an optimal rate" can be established: One need only apply a diagonal procedure to an enumeration of Turing machines to eventually arrive at the optimal device (optimal payoff rate) for any effectively solvable environment. Such procedures are of no interest in studies of adaptation because the corresponding "times of convergence" are in general many orders of magnitude greater than that required by other possible strategies. (Applied to real adaptive systems: such strategies entail extinction.)

It can be shown that near-optimal strategies exist for a class of environments, $\mathcal{E}$, which includes a rich class of normalized game trees (cf. von Neumann, Morgenstern [12]). No systematic study over a structured family of pairs $(\mathcal{E},\mathcal{Q})$ exists and the existence of strictly near-optimal strategies over any interesting pair $(\mathcal{E},\mathcal{Q})$ is still an open question. Studies under way indicate an interesting class of near-optimal strategies results if one produces successive sets of strings (corresponding to recursive definition of $\underline{A}_{d(t)}$) through recombination of fragments of duplicates of extant strings (duplication rate determined by payoff) — a translation into this context of the genetic process of crossover with epistatic effects (see Kimura [7]). It can be shown that the techniques employed by Samuel [10], Friedberg [4], and Bledsoe-Browning [1] (among others) are essentially special cases of this class.

——— * ———

REFERENCES

1. Bledsoe, W. W., and Browning, I., Pattern recognition and reading by machine. Proceedings of the Eastern Joint Computer Conference, 225-232 (1959).
2. Burks, A. W., Computation, behavior and structure in fixed and growing automata. Self-Organizing Systems, 282-311, Pergamon Press (1960).
3. Burks, A. W., and Wright, J. B., Theory of Logical Nets. Proceedings of the Institute for Radio Engineers, 41, 1357-1365 (1953).
4. Friedberg, R. M., A learning machine: part 1. IBM Journal of Research and Development, 2, 1, 2-13 (1958).
5. Holland, J. H., Iterative circuit computers. Proceedings of the Western Joint Computer Conference, 259-265, (1960).
6. Holland, J. H., Universal embedding spaces for automata. forthcoming in a festschrift for Norbert Weiner.
7. Kimura, M., On the change of population fitness by natural selection. Heredity, 12, 145-167 (1958).
8. Kleene, S. C., Representation of events in nerve nets and finite automata. Automata Studies, 3-41, Princeton University Press (1956).
9. McNaughton, R., On nets made up of badly timed elements, I. Summer Conference Notes on Parallel Computers, Automata and Adaptive Systems, The University of Michigan (1962).
10. Samuel, A. L., Some studies in machine learning, using the game of checkers. IBM Journal of Research and Development, 3, 210-229 (1959).
11. von Neumann, J., The theory of automata: construction, reproduction, homogeneity. unpublished manuscript.
12. von Neumann, J., and Morgenstern, O., Theory of Games and Economic Behavior. Princton (1947).

THE UNIVERSITY OF MICHIGAN
COLLEGE OF LITERATURE, SCIENCE, AND THE ARTS
Department of Communication Sciences

Technical Report

ITERATIVE CIRCUIT COMPUTERS:

CHARACTERIZATION AND RÉSUMÉ OF ADVANTAGES AND DISADVANTAGES

John H. Holland

ORA Project 03105

under contract with:

DEPARTMENT OF THE NAVY
OFFICE OF NAVAL RESEARCH
CONTRACT NO. Nonr-1224(21)
WASHINGTON, D.C.

administered through:

OFFICE OF RESEARCH ADMINISTRATION ANN ARBOR

February 1965

RESEARCH PROGRESS REPORT

Title: "Iterative Circuit Computers: Characterization and Résumé of Advantages and Disadvantages," J. H. Holland, University of Michigan Technical Report 03105-35-T, February 1965; Nonr-1224(21).

Background: The Logic of Computers Group of the Communication Sciences Department of The University of Michigan is investigating the application of logic and mathematics to the design of computing automata. Studies of adaptation in an automaton framework forms a part of this investigation.

Condensed Report Contents: This report begins with a revised characterization of the class of iterative circuit computers (i.c.c.), the various members of this class being determined by the substitution instances of a quintuple (A, A^o, X, f, P) where A is any finitely generated abelian group, A^o is a selected subset thereof, X is any finite set, f and P are finite functions over extensions of X. A brief description of the relation of i.c.c.'s to universal embedding spaces follows. The report concludes with a discussion of some of the advantages and disadvantages of i.c.c. organization for computers constructed of microelectric modules.

For Further Information: The complete report is available in the major Navy technical libraries and can be obtained from the Defense Documentation Center. A few copies are available for distribution by the author.

The discussion which follows is divided into two parts: The first part presents a characterization of the various possible organizations for iterative circuit computers — a streamlined version of the 1960 characterization [4]. The second part comments on some advantages and disadvantages of these organizations for practical machines, particularly micromodular arrays.

The class of iterative circuit computers is the set of all devices (automata) specified by the admissible substitution instances of the quintuple (A, A^0, X, f, P). Each particular quintuple designates a distinct iterative circuit computer organization. Intuitively the five parts of the quintuple determine the following features of the organization:

(i) selection of A determines the underlying geometry of the array, particularly the dimension — thus, among other things A determines whether the array is to be planar, 3-dimensional or higher dimensional;

(ii) selection of A^0 determines the standard neighborhood or connection scheme of modules in the array — this A^0 determines the number and arrangement of modules directly connected to a given module;

(iii) selection of X determines the storage register capacity of the module;

(iv) selection of f determines the instruction set and related operational characteristics of the module;

(v) selection of P determines the path-building (addressing) capabilities of the modules — see below.

More formally, the admissible substitution instances of each of the five quantities are:

(i) A must be some finitely generated abelian group having a designated set of generators, say $g_0, g_1, \ldots, g_n$, with the restriction that no constraining relations involve g_o. That is, the group is free on g_o. The positions of modules in the array are indexed by elements of the subgroup A' generated by $g_1, \ldots, g_n$. The time-step is given by the exponent of g_o. Thus $\alpha = g_o^t g_1^{j_1} \ldots g_n^{j_n}$, an element of A, specifies time-step t at the module having coordinates $(j_1, \ldots, j_n)$. By choosing the subgroup A' appropriately the modules can be arranged in a plane, or a torus, or an n-dimensional array, etc. For example, if A' is free on two generators g_1, g_2, an infinite planar array is specified. If the constraining relations

$$g_1^{100} = e$$

$$g_2^{100} = e \quad \text{, where e is the group identity,}$$

are added, a 2-dimensional torus 100 modules in each diameter (10,000 modules total) is specified.

(ii) A^o must be a finite set of elements, $\{a_1, \ldots, a_k\}$, belonging to the subgroup A' of A. For a module at arbitrary location, A^o specifies the arrangement of directly connected modules. Thus the modules directly connected to the module indexed by $\alpha = g_o^t g_1^{j_1} \ldots g_n^{j_n}$ will be the modules at $a_i \alpha = g_o^t g_1^{j_1+k_{i1}} \ldots g_n^{j_n+k_{in}}$, where $a_i = g_1^{k_{i1}} \ldots g_n^{k_{in}}$. For example, if there is a module at (j_1, j_2) relative to generators g_1, g_2, and the directly connected modules are to be at coordinates (j_1+1, j_2), (j_1, j_2+1), (j_1-1, j_2), and (j_1, j_2-1), then A^o should be the set $\{g_1, g_2, g_1^{-1}, g_2^{-1}\}$, where g^{-1} is the group inverse of g.

(iii) X can be an arbitrary finite set. The set of internal states of the module is the set $S = X \times Y$ where Y is the cartesian product $\Pi_{i=1}^{k} Y_i$, $Y_i = \Pi_{j=1}^{k} \{a_j \cup \phi\}$ and $a_j \in A^o$. That is, Y is the set of k x k matrices with entry Y_{ij} being a_j or ϕ. The set X corresponds roughly to the possible states of the module's storage register; the set Y consists of the possible gate configurations for the paths — see the transition equations below.

(iv) f can be an arbitrary finite function of the form

$$f: \{S \cup \phi\}^k \rightarrow S \quad .$$

f deterimines the instruction set, that is, the permissible transitions of the storage register states — see the transition equations.

(v) P can be an arbitrary finite function of the form

$$P: S \rightarrow Y \quad .$$

P determines changes in path gating — see the transition equations.

Having chosen (A, A^o, X, f, P), the behavior of the corresponding iterative circuit computer is completely determined by the following state transition schema:

[S(α) will designate the element of S associated with α under the mapping defined recursively by the transition schema. Under interpretation S(α) designates the internal state of the module with space-time coordinates $(t, j_1, \ldots, j_n)$ corresponding to $\alpha = g_o^t \, g_1^{j_1} \ldots g_n^{j_n}$. This convention will also be used for the components of S and, in particular, $Y_{ij}(\alpha)$ will designate the value of element Y_{ij} in the matrix Y associated with α. Note also, that $g_o\alpha = g_o^{t+1} \, g_1^{j_1} \ldots g_n^{j_n}$ designates the module at the same space coordinates as given by α, but at time t+1 rather than t.]

The transition schema for $Y(g_o\alpha)$ determines the path-gating at time t+1 in the corresponding module in terms of the internal state of the module at time t, $S(\alpha)$. Under interpretation, if $Y_{ij}(\alpha) = a_j$ the gate is open so that information can be passed <u>without</u> a time-step delay <u>from</u> the module at $a_j\alpha$ through the module at α <u>to</u> the module at $a_i^{-1}\alpha$; if $Y_{ij}(\alpha) = \phi$ the gate is closed. In other words $Y(\alpha)$ tells how information is to be channeled through the module to its immediate neighbors; the matrices for these neighbors tell how the information is to be sent on from there, etc. (Details of the information transfer are given by the transition equations for $S(g_o\alpha)$).

$$Y_{ij}(g_o\alpha) = Y_{ij}(\alpha) \text{ if } Q_{ij}(\alpha) = 0 \text{ and } P_{ij}(\alpha) = a_j$$

$$= P_{ij}(\alpha) \text{ otherwise}$$

where $P_{ij}(\alpha)$ is the matrix element (i,j) of $P(S(\alpha))$

and $Q_{ij}(\alpha) = \Lambda_{h=1}^{k} q_{jh}(a_j\alpha)$

$$q_{jh}(\beta) = 0 \text{ if } Y_{jh}(\beta) = \phi \text{ and } P_{jh}(\beta) = a_h$$

$$= 1 \text{ if } P_{jh}(\beta) = \phi$$

$$= \Lambda_{\ell=1}^{k} q_{h\ell}(a_h\alpha) \text{ otherwise}$$

where $\Lambda_{x=1}^{k} q_{jx}$ is the conjunction of the q_{jx}.

Under interpretation $P_{ij}(\alpha)$ specifies a proposed gate-setting for time t+1 at the given module. $Q_{ij}(\alpha)$ prohibits any change in the gate-setting, <u>if</u> there are any changes elsewhere in the path leading through that particular gate. This prohibition prevents the following unstable situations:

(i) a cycle of connections without delay (operation of modules belonging to such a cycle would in general be indeterminant).

(ii) an indefinitely long chain of connections without delay (otherwise a possibility in certain interesting infinite arrays).

The transition equations for $X(g_o\alpha)$ are given in terms of a function I: $B\rightarrow S$, defined for a subset B of A. Under interpretation I represents input to the computer:

$$X(g_o^N\alpha) = f_X(S'(a_1\alpha, 1), \ldots, S'(a_k\alpha, k))$$

where f_X is the restriction of f to X

and $S'(\beta, i) = S(\beta)$ if $Y_i(g_o\beta) = (\phi, \ldots, \phi)$

and $I(\beta)$ is not defined

$= I(\beta)$ if $Y_i(g_o\beta) = (\phi, \ldots, \phi)$ and $I(\beta) \varepsilon S$

$= f(S'(Y_{i1}(\beta)\cdot\beta, 1), \ldots, S'(Y_{ik}(\beta)\cdot\beta, k))$

otherwise

where $\phi 3 = \phi$ and $S'(\phi, j) = \phi$.

Before going on to the practical advantages and disadvantages of such organizations I would like to relate iterative circuit computers to the infinite automaton set down by von Neumann [9]. Von Neumann's cellular automaton is a natural generalization of the McCulloch-Pitts type of automaton (all primitive elements involve a unit delay): it is an infinite 2-dimensional iterative array of an automaton of that type. In an analogous way iterative circuit computers are natural generalizations of the Burks-Wright type of finite automaton (all switches have negligible delay). Kleene proved that any behavior realizable by a finite automaton of the Burks-Wright type (i.e. any regular event) can be realized by a McCulloch-Pitts automaton at the cost of a constant delay of two time-steps [6]. Thus there is a negligible difference behaviorally between these two types of finite automata. This is not true of the corresponding infinite generalizations. There is no way of simulating finite automata in a von Neumann array in such a way that the corresponding behaviors all occur with some constant change of time scale, $t' = kr + c$; the more complex the finite automaton simulated, the slower the simulation. On the other hand, an arbitrary finite automaton can be simulated in an iterative circuit computer, preserving not only behavioral timing ($k = 1$) but also details of local structural and behavioral relations. (E.g., the simulation can reflect differences corresponding to realization of a switching function in terms of $\{\wedge, \vee, \bar{}\}$ vs. realization in terms of the stroke function, $\{|\}$.) In fact, it can be shown such simulation is possible in iterated cellular arrays with locally finite information transfer characteristics only if there is provision for the "making" and "breaking" of non-delay paths. For development of these points the reader is referred to [5].

As a potential organization for computers constructed from micro-electronic components, iterative circuit computers (i.c.c.) exhibit two principal characteristics:

(1) the entire processing part of the computer can be made up of identical modules uniformly interconnected,

(2) within limits of overall storage capacity, the resulting computer can execute an arbitrary number of different programs simultaneously.

The first characteristic offers the following advantages:

(i) set-up costs can be spread over a large number of identical units,

(ii) production, inventory and repair can all be centered on a single integrated component,

(iii) short leads and simple inter-unit connection procedures can be used,

(iv) interface between units can easily be kept standard so that improved units (even those involving new instructions) can be added, thus increasing capacity or capability, without shutdown — programs already written will run correctly on the modified device.

The second characteristic offers the following advantages:

(i) high computation rates for calculations which can be executed in parallel fashion (many outstanding problems owe their computational difficulty to the fact that the underlying process is essentially parallel, forcing single sequence machines into a scanning procedure; in such cases parallel computations offer computation rates unobtainable by any feasible decrease in the cycle time of a single sequence machine).

(ii) space-sharing becomes possible, i.e., the computer can be divided into arbitrary independently-operating sub-computers as required (with store protect features, the sub-computers can be prevented from interaction under central control) and reorganization can be effected, upon demand, under program control.

(iii) because all units are identical and because programs can be executed simultaneously, diagnostic procedures can continually sample modules, with a negligible loss in efficiency — if faulty or failing modules are located, it is possible to "program around" them until replacement.

There are several important disadvantages to i.c.c. organization. Perhaps the most serious disadvantage, to the present time, has been the number of active elements required in each module: several hundred for reasonable flexibility and ease of operation. Closely related, is the low average use factor for some elements in the module. For example, each module usually will have arithmetic capability; yet at any given time only a small fraction of the modules will be using this capability. (It is worth noting that reduced use factors are tolerated even in contemporary single sequence machines as a matter of convenience: compiled programs for simulation of highly parallel systems can be 3 or 4 times slower than "hand-crafted" programs, resulting in a lowered effective use factor for the arithmetic unit.) The importance of these disadvantages decreases as the average cost of a module drops. If the cost of production is largely set-up cost, it may be possible to produce complicated modules for what it presently costs to produce and assemble a few transistors. Should this happen, average use factor for individual elements is no longer a reasonable measure of overall machine efficiency.

Other potential disadvantages center on problems of internal access and broader problems of programming languages and programming convenience. Some aspects of these problems have been investigated (see [2], [7], and [8]), but they remain largely unexplored. The internal accessing problem stems from the procedure used to locate operands — an operand is accessed by "building a path" (opening a sequence of gates) to it. Two difficulties arise. One depends on the number of operands accessible to any module via a reasonable number of path-building operations. The other concerns path interference (crossover, multiple access, etc.). The first of these is less a problem than it seems at first sight. Assume that no more than 10 gates can be opened, for a given path extension, in one machine cycle (i.e. each path can be extended through at most 10 modules in one machine cycle). Still, in two machine cycles, any of 400 modules can be accessed from any given module in a 2-dimensional machine where each module has four nearest neighbors. The problem of path interference is more serious; however there are at least two ways of alleviating this problem: higher dimensionality [8] or path-building via trunk lines [3]. Both can be used together and both also further reduce the first problem.

Programming convenience and the design of programming languages for an i.c.c. are extensive problems not likely to be solved in a fell swoop. One or two comments may show the scope of the problem and where hope lies. For some kinds of problems, programming is actually simpler than for conventional single sequence machines. The solution of partial differential equations in 3-space and time (the weather problem, etc.), by conventional methods on a single-sequence computer, requires that about 90% of the program be given over to scanning logic, boundary manipulation, etc. These considerations are eliminated in an appropriate i.c.c., since the basic grid-point sub-routine can simply be copied throughout — all grid-points are then updated simultaneously. In other contexts, because programs can be executed simultaneously, we face the usual difficulties of parallelism: asynchrony and priority. Certain natural operations, such as association along paths, complicate the problem. Techniques used in the design of asynchronous circuits are useful here. For example, sub-programs can be written so that, when their part of the calculation is ready the corresponding output registers are assigned ready status; sub-programs go to execution status when all operands required are in ready status. An example of an i.c.c. compiler, along somewhat different lines, appears in [1].

There is a great class of highly-parallel problems intrinsically beyond the capability of any presently feasible single sequence machine: the unrestricted weather problem, magnetohydrodynamic problems, command and control problems, simulations of biological and ecological systems, and so on. Iterative circuit computers offer the possibility of treating these problems in parallel fashion — thus making them accessible to computation. On balance I would say that the feasibility of an i.c.c. rests upon resolution of the first-mentioned disadvantage: design and production of a module with several hundred active elements at a relatively low cost. The other problems will almost certainly be resolved sufficiently to permit useful computation if this can be accomplished.

ACKNOWLEDGMENT

The work described here was in part supported by the Office of Naval Research under contract Nonr 1224(21).

REFERENCES

1. Comfort, W. T., "Highly Parallel Machines," Workshop on Computer Organization, ed. Barnum, A. A., and Knapp, Spartan, 1963.
2. Comfort, W. T., "A Modified Holland Machine," Proc. 1963 Fall Joint Computer Conference IEEE, 1963.
3. Gonzalez, R., "A Multi-Layer Iterative Circuit Computer," Transactions on Electronic Computers EC-12, 5, 781-790, 1963.
4. Holland, J. H., "Iterative Circuit Computers," Proc. Western Joint Computer Conference, 259-265, IEEE, 1960.
5. Holland, J. H., "Universal Spaces: A Basis for Studies of Adaptation," to appear in Automata Theory, ed. Caianiello, E. R., Academic Press, 1965.
6. Kleene, S. C., "Representation of Events in Nerve Nets and Finite Automata," Automata Studies, 3-41, Princeton, 1956.
7. Newell, A., "On Programming a Highly Parallel Machine to be an Intelligent Technician," Proc. Western Joint Computer Conference, 267-282, IEEE, 1960.
8. Squire, J. S., and Palais, S. M., "Programming and Design Considerations of a Highly Parallel Computer," Proc. Spring Joint Computer Conference, IEEE, 1964.
9. von Neumann, J., "The Theory of Automata: Construction, Reproduction, Homogeneity, unpub. manuscript.

REPRINTED FROM

COMPUTER AND INFORMATION SCIENCES — II

ACADEMIC PRESS INC., NEW YORK

Nonlinear Environments Permitting Efficient Adaptation

John H. Holland

UNIVERSITY OF MICHIGAN
ANN ARBOR, MICHIGAN

The study of adaptation is related to the study of triples of the form $\langle \mathscr{E}, T, \chi \rangle$: $\mathscr{E}$ is a class of environments (say a set of finite automata each with an associated payoff function μ over its states) representing what is unknown to the adaptive system; T is a set of adaptive strategies, each of which is a technique for selecting among alternative organizations on the basis of information received from the environment (e.g., a finite automaton is modified on the basis of inputs from $E \in \mathscr{E}$); χ is a criterion for ranking the elements of T according to performance over $\mathscr{E}$. An environment is linear over a decomposition of environmental states, $S_E = \prod_{i=1}^{m} S_{Ei}$, if there exists μ_i such that for all $s \in S_E$ payoff $\mu(S_{h_1}, \ldots . S_{h_m}) = \sum_i \mu_i(S_{h_i})$; an environment is nonlinear otherwise.

An explicit expression is given for the performance of an interesting class of adaptive strategies over any stationary nonlinear environment. In this expression control parameters are coupled with various orders of nonlinearity. With its help one can prove that particular strategies $\tau \in T$ (which can be constructively given) are universal over a wide range of nonlinear environments $\mathscr{E}$ in the sense that each meets a criterion χ akin to avoidance of gambler's ruin over the whole range $\mathscr{E}$ (even when compared to a strategy which takes optimal action from the first instant against whatever $E \in \mathscr{E}$ is presented).

I. Introduction

What kind of organization enables a system to solve problems foreseeable only in vague outline? How can a system organize itself to detect and act upon environmental regularities? How do environmental regularities affect a system's rates of information processing and reorganization? Many studies of natural and artificial systems come, sooner or later, to such questions. In several specific contexts—genetics, morphogenesis, psychology, control theory—the processes eliciting these questions come under the heading "adaptation." As we shall see, there is in automata theory a common framework for the description of both these processes and a wide range of similar processes occurring in other contexts. This common framework suggests that "adaptation" be used in a wider sense to cover all such processes. Indeed it suggests a good deal more: the possibility of extracting and studying features of adaptation common to all its various guises. The present paper explores this possibility.

Much useful information about adaptation has been obtained by simulation and study of formal models of checker-playing, theorem-proving, problem-solving, and pattern recognition (e.g. Bledsoe [3], Friedberg [6], Gelernter [7], Minsky [13], Newell [17], Samuel [19], Selfridge [20], Steck [21]); different but equally interesting results have been obtained by a similar approach to neural nets and homeostatic mechanisms (e.g. Ashby [1], Milner [12], Rochester [18], Turing [23]); and there is a significant statistical theory of the changes in fitness of genetic systems under various mating rules, mutation rates, and population pressures (e.g. Fisher [4], Fraser [5], Kimura [10], Moran [14]). At the same time, analogies drawn from automata theory and information theory have been helpful in closely allied areas, e.g., the "coding" problems of biochemical genetics, the formal study of self-reproduction and adaptive control (Bellman [2], Goldberg [8], von Neumann [15]). And underlying all is a great body of empirical knowledge about adaptation in biological systems (recently summarized masterfully in Mayr [11], see also Grant [9], Tax [22]).

Despite many insights produced by these and related approaches, we are still a long way from a general understanding of adaptive mechanisms. Useful and suggestive results often remain in comparative isolation. Just as different aspects of neural action are exaggerated in the giant axon of the squid, the paired acoustic cells of the moth ear, and the neurons generating the scratch-reflex of the dog, so it is with various adaptive systems. One aspect of adaptation will be prominent in one system, obscured in another. The task of theory is to ferret out general principles from this complex tangle of problems, studies, and data. At the outset we gain in sophistication by describing the different systems in a common formal framework. (It will be the task of Sec. II to back this claim.) Beyond this, a theoretical approach is essential if the problems of adaptation are not to be tackled anew, and in piecemeal fashion, each time they arise in a new context.

The theoretical approach taken here will start from a careful look at the notion of a "universal adaptive strategy." I should say at once that the crucial question is not: Does there exist an adaptive strategy which always attains the optimum possible behavior in some "universe" of situations? It is easy to design a system which will try, in some order, all computable procedures or all organizations which can be effectively defined; the "universe" in this case could be very extensive indeed, and yet the system would eventually attain the optimal organization in any situation drawn from the universe. The crucial questions center on ranking various adaptive strategies according to their efficiency or fitness over the range of possible environmental situations. Can an adaptive strategy perform *well* in a reasonably extensive variety of situations? Does a given adaptive strategy make *better* use than another of the information it receives? A careful formulation of this kind of question requires some effort. The formulation given here will be developed until, in Sec. IV, it will be possible to estimate the rate of adaptation of various kinds of adaptive strategies over an extensive class of nonlinear environments. A corollary indicates the existence of adaptive strategies which perform well (in a rather strong sense) when confronted with any element of the class. The result is constructive, indicating the form of the control parameters and the techniques the system can employ to adjust the parameters automatically as information accumulates. Much that is known about particular adaptive systems can be reinterpreted in these terms. For instance, one can readily represent the effects upon adaptive efficiency of genetic mechanisms such as mutation, gene duplication plus insertion, deletion, and translocation augmented by crossover and recombination. Using exactly the same procedures, one can infer both the efficiency of a basic scheme like Samuel's, over a variety of games and opponents, and lines along which improvements can be effected. Section IV will be approached over familiar ground: Sec. III will try the framework of Sec. II on adaptive strategies employing linear operators against linear environments.

II. A Formal Framework

Adaptation is a meaningful concept only on the assumption of an environment initially unknown to the adaptive system in one or more of its characteristics. A characteristic can be unknown, from the viewpoint of the adaptive system, only if alternatives are possible. Each distinct combination of alternatives in effect is a distinct environment that the system may possibly face; the set of all possibilities constitutes the range of environments the system must be prepared to face. In what follows this will be formalized by designating a class $\mathscr{E}$ of admissible (or possible) environments.

In a similar vein, it is necessary to specify the techniques, models, or devices available to the adaptive system. The class of devices which the system can bring to bear through rearrangement of its structure will be designated $\mathscr{A}$. An adaptive strategy, then, will correspond to a (conditional) trajectory through the set $\mathscr{A}$, the elements of the set $\mathscr{A}$ being tried in an order determined by the outcome of previous trials. That is, the strategy will dictate successive reorganizations of the adaptive system (the trajectory) according to the information (inputs) the system receives from the environment, $E \in \mathscr{E}$, confronting it.

In general we will be interested in comparing different adaptive strategies confronted with different admissible environments. More formally, if T denotes the trajectories of interest and χ is a given "fitness" criterion, the object is to determine the ordering on T over $\mathscr{E}$ induced by χ. The *sine qua non* of a theory of adaptation is a ranking of adaptive strategies according to "fitness" over the admissible environments. And here we come upon a series of difficulties. Whatever "fitness of an adaptive strategy" is to mean, it is certain to depend upon some criterion of optimality as applied to devices in $\mathscr{A}$ confronted by a specific environment E. The diversity of possible criteria creates the first difficulty. Even if we consider a single environment E, as in control theory, different criteria lead to different solutions. In greater detail, $\mathscr{A}$ corresponds to the admissible controls or control region, E to the controlled process or plant, the criterion χ depends upon the specific control problem, and different criteria implicitly define different elements of $\mathscr{A}$ as optimal. More generally, the criterion appropriate to a given study of adaptation may be maximization of accumulated utility, attainment of a target state in minimal time, minimization of cumulative error, escape from gambler's ruin, maximization of fitness (in Fisher's sense) or survival. And there are others. Somehow this embarrassment of choices must be resolved if we are to have a uniform theory.

Interpretation of the different optimality criteria in terms of admissible devices and environments serves to reduce this embarrassment. It will be easier to see this if we carry out a few preliminaries. Let us begin by restricting our attention for a moment to games, as prototypes of a wide range of environments. We know that every game can be represented by a tree of moves (cf. von Neumann [16]) with payoff assigned to the terminations. Allowance can be made for repeated plays of the game by connecting every termination to the initial vertex ("resetting the game for the next play")—any such graph is strongly connected in the sense that given any vertex there is a directed path from it to any other vertex. It is a natural extension to let $\mathscr{E}$ consist of strongly connected finite automata each with an associated payoff function μ assigning to each vertex a real number (the associated payoff). Next, let the devices $A \in \mathscr{A}$ be finite automata and let A be connected to the environment $E \in \mathscr{E}$ confronting it so that each output symbol of A determines an input symbol of E and vice versa. Thus, when connected to any $E \in \mathscr{E}$, each A determines a unique transition ("move") in E at each instant of time t. That is, A when connected to E determines a sequence of environmental states $e(1)$, $e(2)$, ..., $e(t)$, ..., and hence a payoff sequence $\mu(e(1))$, $\mu(e(2))$, ..., $\mu(e(t))$, ... For convenience rename elements of this payoff sequence $o_{A,E}(t)$. With the help of these preliminaries we can begin the discussion of μptimality criteria.

Given $E \in \mathscr{E}$ and a time $\mathscr{T}$ we can assign to each device A the number

$\sum_{t=1}^{\mathcal{T}} \mu_{A,E}(t)$, the payoff accumulated to time $\mathcal{T}$ when A confronts E. Under the criterion "maximization of accumulated utility (payoff)," one could call A_0 $(E, \mathcal{T})$-optimal with respect to environment E at time $\mathcal{T}$ if, for all $A \in \mathcal{A}$,

$$\sum_{t=1}^{\mathcal{T}} \mu_{A_0,E}(t) > \sum_{t=1}^{\mathcal{T}} \mu_{A,E}(t).$$

It is at once clear that there may be distinct $(E, \mathcal{T})$-optimal strategies for different $\mathcal{T}$. However, given any finite automaton with associated payoff, it is easy to establish that maximum mean payoff per time-step will be attained by cycling around a given closed path in the transition graph for E. If there are elements of $\mathcal{A}$ which can "force" E to cycle on this closed path, then one of them will be $(E, \mathcal{T})$-optimal for all $\mathcal{T}$ greater than some $\mathcal{T}_{\min}$. If μ assigns 1 to states on the closed path and 0 to states not on it, the optimal device will be the one which most quickly forces E to the optimal closed path. Here we make contact with a criterion used to define optimal control processes, "attainment of the target state in minimal time." "Target state" is simply a special case of "optimal closed path." Thus, with μ as given, devices optimal under the "maximal accumulation" criterion are also optimal under the "minimal time" criterion. By using a payoff function which assigns to the target states of E a value zero, and to all other states a value equal to the corresponding error, another familiar criterion, "minimization of cumulative error," becomes a minimization criterion equivalent to the "maximal accumulation" criterion.

On the same basis, "escape from gambler's ruin" is essentially a weaker version of the "maximal accumulation" criterion. Instead of requiring, for all $A \in \mathcal{A}$,

$$\sum_{t=1}^{\mathcal{T}} \mu_{A_0,E}(t) > \sum_{t=1}^{\mathcal{T}} \mu_{A,E}(t)$$

it is required that

$$\sum_{t=1}^{\mathcal{T}} \mu_{A_0,E}(t) > c \sum_{t=1}^{\mathcal{T}} \mu_{A,E}(t)$$

for some $c > 0$. (This is a modified version: instead of treating A_0 and A as opponents in a zero sum game, each is matched with a common opponent E and "winnings" are compared.) This weaker criterion, of course, enlarges the set of "optimal" devices for any given E and $\mathcal{T}$. As we shall see, it can usefully be extended to cases where E possess an infinite number of states if, for some A_0 and all $A \in \mathcal{A}$,

$$\lim_{\mathcal{T} \to \infty} \left[\sum_{t=1}^{\mathcal{T}} \mu_{A_0,E}(t) \Big/ \sum_{t=1}^{\mathcal{T}} \mu_{A,E}(t) \right]$$

is defined and greater than zero. The result is a still weaker condition on A_0 with dependence on $\mathcal{T}$ removed. Two other criteria of optimality, Fisher's "fitness" and "survival" are best considered in relation to trajectories in $\mathcal{A}$; their discussion will be postponed while the topic of trajectories is reintroduced.

Each of the foregoing optimality criteria is distinct enough from its fellows to be appropriate to systems for which one or more of the others are inappropriate. However, the considerations just given suggest the payoff accumulation functions as the common basis underlying the several criteria. A theory of adaptation based on these functions should be interpretable and useful in a wide range of particular applications. Granting this, there still remains a formidable set of difficulties. Up to now, we have only considered the ranking of particular devices acting on particular environments. The corresponding criteria have been "local," applying only to individual pairs (A, E) and not to trajectories through $\mathcal{A}$. To obtain information about trajectories as strategies for adaptation it is necessary to rank the trajectories over the full set of possible environments $\mathcal{E}$.

An example will emphasize this necessity: Consider a trajectory τ_0 which selects elements of $\mathcal{A}$ in an order independent of the particular element of $\mathcal{E}$ confronting it. Suppose the first element selected by τ_0 is optimal with respect to $E_0 \in \mathcal{E}$. If E_0 is indeed the environment confronting τ_0, τ_0 will certainly accumulate payoff at the maximum possible rate. But, given almost any other $E \in \mathcal{E}$, τ_0 will perform quite poorly assuming $\mathcal{E}$ and $\mathcal{A}$ are nontrivial—τ_0 produces elements of $\mathcal{A}$ in a fixed order, irrespective of information received from E, and thus will almost always require a very long time to stumble upon the optimal device corresponding to E (cf. Samuel's [19] comments on the enumeration of alternatives in the relatively simple environment of checkers).

This example points up both a difficulty and a factor important to its resolution. In the study of adaptation, we are interested in strategies which select successive devices from $\mathcal{A}$ in a conditional fashion, conditional upon information sampled from the particular element of $\mathcal{E}$ present. No such strategy can be "universal" *if* the condition for universality is that, given any $E \in \mathcal{E}$, the strategy must at all times accumulate payoff at least as rapidly as any other admissible strategy. This failure follows from the fact that any adaptive strategy τ samples only a finite number of devices up to time t_0. If $\mathcal{E}$ is at all interesting, there will be some environment $E_0 \in \mathcal{E}$ for which none of the devices sampled by τ to time t_0 is optimal. Given τ_0 as before, it will initially accumulate payoff at a higher rate than τ when confronted with E_0. But, under the foregoing criterion, this eliminates τ as a candidate for "universality." Thus, if the notion of "universality" is to be useful to us, it must be based on a weaker criterion. On the other hand the criterion must not be so weak that it fails to eliminate the unconditional strategies (which must do poorly over almost all of $\mathscr{E}$). We would like to examine the relative capabilities of various conditional strategies while eliminating unconditional or enumerative strategies as unfit.

With this in mind, we return to the pair of criteria more directly concerned with trajectories. Fisher's [4] "fitness" is a measure of the expected number of offspring of an organism or, more precisely, of particular gene combinations. To accommodate Fisher's notion, the definition of adaptive strategy must be extended to include trajectories through *sets* of devices drawn from $\mathscr{A}$ (alternatively, $\mathscr{A}$ can be redefined so that its elements are sets of devices). Each device in $\mathscr{A}$ then corresponds to a particular gene combination and the strategy simply determines successive "populations" of such devices. The set of devices operative at a given time can be thought of as a set of trials or samples of the environment's payoff function. The results of these trials are used to determine the number and kinds of the next set of trials. That is, the adaptive strategy determines the new set of devices from the successes and failures of the current set. Payoff $\mu_{\tau,E}(t)$ is measured in units of offspring and the contribution of the parent device to the set of offspring—its fitness—is determined by the payoff it achieves. Different adaptive strategies under this interpretation correspond to different rules of mating, crossover, etc. In keeping with Fisher's treatment, we can then assign to any strategy τ at time $\mathscr{T}$ the total offspring produced to that time, $\sum_{t=1}^{\mathscr{T}} \mu_{\tau,E}(t)$. Comparison on this basis is, then, an extension of the "maximal accumulation" criterion to trajectories. "Survival" is a weakened version of this criterion, requiring only that the strategy not cease to produce offspring. Making an interpretation similar to the one for "gambler's ruin," this becomes a requirement that the accumulated offspring under τ_0 not become vanishingly small in comparison with the accumulation under some other strategy τ. More formally, for all τ,

$$\lim_{\mathscr{T}\to\infty}\left[\sum_{t=1}^{\mathscr{T}} \mu_{\tau_0,E}(t) \Big/ \sum_{t=1}^{\mathscr{T}} \mu_{\tau,E}(t)\right] > 0$$

an extension of the gambler's ruin criterion to trajectories.

We see from this and the earlier discussion that the payoff accumulation functions appear centrally in the interpretation of each of the six criteria considered. The list could be extended, but to little present purpose. This discussion aims only at giving plausibility to a particular approach to the ranking of adaptive strategies: the study of the payoff accumulation by different trajectories in $\mathscr{A}/\mathscr{E}$ complexes as means of obtaining widely interpretable results.

To summarize then: This paper is concerned with adaptive systems which can be studied within the framework provided by triples of the form $\langle \mathscr{E}, T, \chi \rangle$. $\mathscr{E}$ is the set of possible environments represented, for present purposes, by a set of strongly connected finite automata with associated payoff function μ from states to payoffs. T is the set of adaptive strategies of interest, represented by a set of conditional trajectories over a set $\mathscr{A}$ of devices represented, again for present purposes, by a set of finite automata. The device $A \in \mathscr{A}$, corresponding to a given point in the trajectory $\tau \in T$, will be so specified that its output symbols correspond to input symbols of the $E \in \mathscr{E}$ confronting τ; the output symbols of E will similarly be mapped onto the input symbols of A. The successor of device A in τ will be conditional upon the input recieved from E.

The object of the study is to determine (aspects of) the ranking on elements of T induced by the given criterion function χ. The burden of the discussion of this section has been that the rankings imposed by a wide range of criteria can be usefully studied via the payoff accumulation functions $\sum_{t=1}^{\mathscr{T}} \mu_{\tau,E}(t)$.

We cannot expect to find a single trajectory which yields "maximal accumulation" over all of any nontrivial $\mathscr{E}$—enumerative strategies of the type mentioned earlier will always outrank all other strategies on particular environments. Thus we must look to the weaker "survival" criterion. The hope here is that the criterion is satisfiable but not by enumerative or other intuitively inefficient trajectories. If the criterion is to be serviceable, it should enable us, in the proper context, to derive characteristic structural properties of strategies satisfying it. This seems a lot to ask, but surprisingly we shall see that for some very general and interesting $\mathscr{E}$, the criterion serves just these purposes.

III. Linear Operators and Linear Environments

Some of the concepts and interpretations needed later are developed here using the framework $\langle \mathscr{E}, T, \chi \rangle$ of Sec. II on relatively familiar ground: $\mathscr{A}$ will consist of techniques corresponding to the linear operator scheme of Samuel's [19] checker-playing program; T will consist of various schemes τ for modifying the operators; and $\mathscr{E}$ will consist of games with linear payoff functions. We will begin by defining the linear operator techniques and linear payoff functions. Then these will be related to the payoff accumulation functions of Sec. II.

The techniques in $\mathscr{A}$ will have as a basis a fixed set of functions $\{0_j\}, j = 1, \ldots, m$, where each 0_j assigns a real number to each environmental (output) state (e.g., in Samuel's case, the environmental state is the current board configuration and a particular 0_j determines some feature of the configuration such as difference in number of pieces, average penetration of pieces beyond the center line, etc.). A particular $A_i \in \mathscr{A}$ is determined by an m-tuple of weights $(a_{i1}, \ldots, a_{im})$ used to form the linear operator $\sum_{j=1}^{m} a_{ij} 0_j$, where, for convenience, it will be assumed that the admissible weights at each position are drawn from a single set $\{\sigma\}$. The linear operator A_i is used (directly or indirectly) to assign values to the alternative outputs (the configurations which can be reached by given moves). The output with highest assigned value is then transmitted to the environment (the board configuration is modified accordingly and the next iteration of the process is initiated).

Now let $E \in \mathscr{E}$ be a two-person game with a fixed pure strategy assigned to the opponent. Brief reflection shows that any technique A_i, when confronted with E, will inevitably come to the same termination of the game tree and be payed off a fixed amount accordingly. Each m-tuple $(a_{i1}, \ldots, a_{im})$ in effect names a particular termination and receives payoff accordingly. But then the payoff function which assigns payoff to the terminal states of E can be rewritten as a function assigning payoff to each m-tuple, $\mu_E \colon \{(a_{i1}, \ldots, a_{im})\} \rightarrow R$. The set of possible environments $\mathscr{E}$ arises naturally if there is a variety of opponents (employing various pure strategies, perhaps including an opponent playing minimax) or a variety of games or both. In these cases μ_E can be rewritten as indicated for each $E \in \mathscr{E}$. The same discussion could be extended to mixed strategies and beyond, but this example will serve present purposes.

An environment E will be called linear with respect to the basis functions $\{\theta_j\}$ if, for all $A_i \in \mathscr{A}$,

$$\mu_E(a_{i1}, \ldots, a_{im}) = \sum_j \mu_j(a_{ij}) \tag{3.1}$$

for some set of functions $\{\mu_j\}$ each of which maps the set of admissible weights $\{\sigma\}$ into the reals,

$$\mu_j \colon \ \{\sigma\} \rightarrow R.$$

With μ_E rewritten to apply to m-tuples of $\mathscr{A}$, we see that selection of a linear operator constitutes a sampling of the value of μ_E for the corresponding argument. In these terms the object of the adaptive system is to sample the function μ_E efficiently so as to rapidly locate arguments (elements of $\mathscr{A}$) with high payoff. Useful contact can be made with mathematical genetics by treating this sampling of elements of $\mathscr{A}$ as a parallel process. That is, following the latter part of Sec. II, payoff is measured in units of offspring; thus, if several operators are tested against an environment E at time t, the outcome determines the number of new operators that can be formed and tested at time $t + 1$. (In a moment we shall see that measuring payoff in terms of offspring is convenient but not essential.) Let us consider a specific parallel strategy.

(1) A finite set of weights $\{\sigma_1, \ldots, \sigma_k\}$ will be used for the linear operators. At time t a set of $n(t)$ operators, $\mathscr{A}(t) = \{A_1(t), A_2(t), \ldots, A_{n(t)}(t)\}$ will be tested against $E \in \mathscr{E}$. Let the weights specifying $A_i(t)$ be $[a_{i1}(t), a_{i2}(t), \ldots, a_{im}(t)]$. Note that the number of occurrences of weight σ_h at position j in the set $\mathscr{A}(t)$

is given by

$$n_{jh}(t) = \sum_{i=1}^{n(t)} \delta_h(a_{ij}(t))$$

where $\delta_h(\sigma_l)$ is 1 if $l = h$, and is 0 if $l \neq h$.

(2) Each $A_i(t) \in \mathscr{A}(t)$ is tested against E and each weight $a_{ij}(t)$ is copied $\mu_E(a_{i1}(t), \ldots, a_{im}(t))$ times. Thus, after the full set $\mathscr{A}(t)$ is tested and copied according to payoff, there will be

$$\sum_{i=1}^{n(t)} \delta_h[a_{ij}(t)]\mu[a_{i1}(t), \ldots, a_{im}(t)]$$

copies of weight σ_h at position j.

(3) Let all copies of the weights associated with position j at the end of step (2) be collected in a single set $W_j(t)$ (which will thus typically contain many duplicates of each weight).

(4) A new operator $A_1(t+1)$ given by $[a_{11}(t+1), \ldots, a_{1m}(t+1)]$ is formed by drawing weight $a_{11}(t+1)$ at random and without replacement from set $W_1(t)$, drawing $a_{12}(t+1)$ from $W_2(t)$, $\ldots$, drawing $a_{1m}(t+1)$ from $W_m(t)$. This process is repeated, forming $A_2(t+1)$, etc., until the sets $W_j(t)$ are exhausted. The number of new operators so-formed is given by

$$\sum_{i=1}^{n(t)} \mu[a_{i1}(t), \ldots, a_{im}(t)]$$

and the new trial set $\mathscr{A}(t+1) = \{A_1(t+1), \ldots, A_{n(t+1)}(t+1)\}$ results.

(5) Return to step (2).

Although this strategy has been specified in discrete terms it can be viewed as a sampling procedure where the probability of trying a particular operator is biased by the outcome of the previous trial: Thus the probability that a given element of $\mathscr{A}(t)$ is the m-tuple $(\sigma_{h_1}, \sigma_{h_2}, \ldots, \sigma_{h_m})$ is given by

$$P(h_1, \ldots, h_m, t) = \prod_{j=1}^{m} P_{jh_j}(t) \tag{3.2}$$

where $P_{jh_j}(t) = n_{jh_j}(t)/n(t)$.

The state of the adaptive system at any time, then, can be looked upon as given by a probability distribution $\psi_{\mathscr{A}(t)}$ over the set $\mathscr{A}$ and the above procedure (1)–(5) as a means of forming $\psi_{\mathscr{A}(t+1)}$ on the basis of $n(t)$ samples drawn from $\psi_{\mathscr{A}(t)}$. We pursue this general line in Sec. IV; for the moment it suffices to note that the above strategy can be looked upon as a means of biasing the sampling of $\mathscr{A}$ on the basis of previous samples—on this interpretation, measuring payoff in terms of offspring is simply a convenient means of attaining the successive distributions $\psi_{\mathscr{A}(1)}, \psi_{\mathscr{A}(2)}, \ldots$.

Using the probabilistic point of view, let us examine the behavior of this strategy when confronted with an environment which is linear (with respect to $\{\theta_j\}$). The expected proportion of weight σ_l at position j at time $t+1$ is given by

$$P_{jl}(t+1) = \frac{\sum_{h_2,\ldots,h_m} P(l, h_2, \ldots, h_m)\mu_E(\sigma_l, \sigma_{h_2}, \ldots, \sigma_{h_m})}{\sum_{h_1,h_2,\ldots,h_m} P(h_1, \ldots, h_m)\mu_E(\sigma_{h_1}, \ldots, \sigma_{h_m})}. \tag{3.3}$$

Two assumptions will enable an analytic treatment:

(i) Assume that negligible error arises in treating expectations like the one just given as actually achieved by the process at each instant (if $\mathscr{A}(t)$ contains enough elements this condition will be satisfied).

(ii) Assume the process to be continuous (so that equations hold for real t and derivatives can be taken).

On this basis, Eqs. (3.2) and (3.1), characterizing the strategy and the linear environment respectively, can be used with (3.3) to show that

$$P_{jl}(t) = \frac{P_{jl}(0)\exp[\mu_j(\sigma_l)t]}{\sum_h P_{jh}(0)\exp[\mu_j(\sigma_h)t]} \tag{3.4}$$

for such strategy-environment combinations. Let $\sigma_{j\bullet}$ be the weight for which $\mu_j(\sigma_{j\bullet}) = \max_h\{\mu_j(\sigma_h)\}$ and let $P_{j\bullet}(t)$ be the corresponding probability. It is easily established from the foregoing equation that $P_{j\bullet}(t)$ approaches 1 exponentially. Thus the approach to the maximum payoff rate per trial, $\mu_* = \sum_{j=1}^{m} \mu_j(\sigma_{j\bullet})$, is bounded below by an exponential function.

The ratio $\rho(t) = n(t)/n_{\max}(t)$ is informative here. As given earlier, $n(t)$ is the total payoff available to the given adaptive strategy at time t. (Recall that this corresponds to the number of trials the system can take at time $t+1$.) $n_{\max}(t)$ is the maximum possible total payoff at time t—this level is achieved only if a system chooses a termination with payoff μ_* on the first trial and plays that termination from then onward. [In the discrete case $n_{\max}(t) = n(0)\mu_*^{\,t}$; in the continuous case $n_{\max}(t) = n(0)\exp(\mu_* t)$.]

Let us consider, first, $\rho(t)$ for an enumerative strategy, τ_{enum}, say uniform random trials without replacement. If only one m-tuple yields payoff μ_*, the expected number of trials to discover μ_* will be $\frac{1}{2}k^m$. The expected payoff for each trial, until μ_* is discovered, will be

$$\bar{\mu} = k^{-m} \sum_{h_1,\ldots,h_m} \mu(\sigma_{h_1}, \ldots, \sigma_{h_m}).$$

Thus the expected minimum of $\rho(t)$ for this strategy will be

$$E(\rho_{\text{min,enum}}) = (\bar{\mu}/\mu_*)\ln[k^m/2n(0)].$$

For example, if $m = 10$, $k = 10$, $\bar{\mu} = \frac{1}{2}\mu_*$, we have

$$E(\rho_{\min,\text{enum}}) \simeq 2n(0) \cdot 10^{-10}.$$

Now let us consider $\rho(t)$ for the best possible search τ_{opt}, when it is known that all environments in $\mathscr{E}$ are linear with respect to $\{0_j\}$. Because of the linearity, an appropriate selection of m-tuples will enable the optimum to be located with just k trials. As a result

$$E(\rho_{\min,\text{opt}}) \simeq (\bar{\mu}/\mu_*) \ln[k/n(0)].$$

If, again, $m = 10$, $k = 10$, $\bar{\mu} = \frac{1}{2}\mu_*$, we have

$$E(\rho_{\min,\text{opt}}) \simeq \tfrac{1}{10}n(0).$$

The parallel strategy defined earlier, because of its exponential approach to the optimum for linear environments, will have a $\rho_{\min}$ of the same order in general as $E(\rho_{\min,\text{opt}})$—discussion is postponed to Sec. IV where this strategy can be handled as a special case.

We have here an indication, simple though it be, of the usefulness of the ratio $\rho(t)$. $E(\rho_{\min,\text{enum}})$ is not just orders of magnitude smaller than $E(\rho_{\min,\text{opt}})$; the difference comes close to being qualitative in that it is not easy to compare the two numbers. Herein lies hope for partially ordering adaptive strategies by using the associated ratios $\rho_{\min}$. At least it seems that enumerative strategies will be tagged with a very low rank by this criterion. We shall see, in Sec. IV, that $\rho(t)$ lends itself to systematic investigation, particularly via study of functions which bound $\rho(t)$. We shall also see that, from knowledge of bounds on $\rho_{\min}$ over a class of environments $\mathscr{E}$, much can be inferred about the corresponding adaptive strategy.

IV. Nonlinear Environments

Before considering nonlinear environments, let us take a closer look at the measure of performance $\rho(t)$ introduced at the end of the last section. To be explicit, let $n_{\tau,E}(t)$ be the payoff available to strategy τ at time t when confronted with environment E; then, generalizing the earlier discussion, let

$$\rho_{\tau_1,\tau_2,E}(t) = \frac{n_{\tau_1,E}(t)}{n_{\tau_2,E}(t)}$$

be the basis for comparing τ_1 and τ_2 over E. In keeping with that discussion, the number $\text{glb}_t\, \rho_{\tau_1,\tau_2 E}(t)$, or bounds thereon, should be of considerable interest. τ_1 will be called *b-optimal* with respect to a strategy τ_2 over environment E if $\text{glb}_t\, \rho_{\tau_1,\tau_2,E}(t) \geqq b$. If τ_1 satisfies this criterion with respect to all

$\tau \in T$ and all $E \in \mathscr{E}$, that is if

$$\operatorname*{glb}_{\tau \in T, E \in \mathscr{E}} \operatorname*{glb}_{t} \rho_{\tau_1, \tau, E}(t) \geq b,$$

then τ_1 will be called $(T, \mathscr{E})$ *b-optimal.* This condition can be interpreted quite directly as assuring that τ_1 will escape gambler's ruin in the sense that the payoff available to τ_1 when compared with that of any $\tau \in T$ in any $E \in \mathscr{E}$ will never be in a ratio less than $b > 0$. In an evolutionary context the condition can be interpreted as saying that the system τ will survive as a proportion of at least b of a population of other systems evolving at the same time. By associating with each $\tau \in T$ the largest b for which τ is $(T, \mathscr{E})$ b-optimal and ordering the τ according to the natural ordering of these associated fractions, we obtain a complete ordering of the elements of T. The discussion of Sec. II makes it plausible that information obtained in determining this ordering will enable determination of orderings imposed by other familiar criteria.

Let us restrict our attention now to classes of environments $\mathscr{E}$ wherein, for each $E \in \mathscr{E}$, there is a maximum payoff rate (per sample) r_{*E}. Let $\zeta(t)$ be an analytic function of t such that the expected payoff rate of strategy τ at time t, when applied to $E \in \mathscr{E}$, is bounded below by

$$r_{\tau,E}(t) = (1 - \zeta(t))r_{*E}. \tag{4.1}$$

It can easily be shown that τ is at least $(T, \mathscr{E})$ b-optimal with respect to any class of strategies T, even if the strategy which achieves rate r_{*E} from $t = 0$ onward belongs to T, if

$$\prod_{t=0}^{\infty} [1 - \zeta(t)] \geq b. \tag{4.2}$$

Whenever τ satisfies the condition implicit in Eq. (4.2), it will be called *strictly* $(\mathscr{E})$ *b-optimal.*

Equation (4.1) can be coupled with Kimura's [10] development of Fisher's equation for rate of change of fitness to yield a condition for nonlinear environmental classes to admit of strictly b-optimal strategies:

The most direct way of accomplishing this, and of bringing Kimura's equation into the present context, is to represent each admissible device in $\mathscr{A}$ as a string over some finite alphabet $\{\sigma_1, \ldots, \sigma_k\}$. In Sec. III the devices in $\mathscr{A}$ were linear operators represented as strings of length m (m-tuples) over a set of k weights. However, with some care, one can obtain string representations, which are reasonably flexible as far as substitution properties, reorganization, etc., for quite general classes of devices (e.g. the class of finite automata). Kimura's equation concerns fitness, but by measuring payoff in units of offspring (or producing a direct iteration for modifying a probability distribution $\psi_{\mathscr{A}}$ over the devices), it becomes an equation for $r_{\tau,E}(t)$:

$$\frac{dr_{\tau,E}}{dt} = \sum_{x \in X} \dot{P}_x \lambda_x + \sum_{\gamma \in \Gamma} P_\gamma \varepsilon_\gamma \frac{d \ln \theta_\gamma}{dt} \tag{4.3}$$

where $x \in X = \{(i, j) \ni i = 1, 2, 3, \ldots, \text{and } j = 1, 2, \ldots, k\}$

$\gamma \in \Gamma = \{x_1 x_2 \cdots x_m \ni x_h \in X, h = 1, \ldots, m \text{ and } m = 1, 2, 3, \ldots\}$

$P_x(t)$ gives the probability that letter σ_j occurs in position i of the strings $\mathscr{A}(t)$ sampled at time t.

λ_x is a lms-estimator, under the distribution $P_x(t)$, of the payoff to be expected from a string having letter σ_j at position i.

$P_\gamma(t)$ gives the probability of the joint occurrence $x_1, x_2, \ldots, x_m$ (letter σ_{j_1} at position i_1, letter σ_{j_2} at position i_2, etc.) amongst the strings $\mathscr{A}(t)$ sampled at time t.

ε_γ is the "error" of a lms-estimate of the expected payoff associated with the joint occurrence $x_1, x_2, \ldots, x_m$, when the probability of the occurrence is P_γ ($\varepsilon_\gamma \neq 0$ indicates an environmental nonlinearity).

θ_γ is the ratio of P_γ to the product of the probabilities of its constituents (i.e. $\theta_\gamma = 1$ if γ is obtained by independent selection of its constituents, otherwise $\theta_\gamma \neq 1$; the θ_γ are, in the present interpretation, control parameters available to the adaptive strategy τ).

Equation (4.3) is an identity holding for any environment and adaptive strategy which can be presented in the above terms. Note that in the equation, the parameters $\{\theta_\gamma\}$ characterizing the adaptive strategy are nicely coupled to the parameters $\{\varepsilon_\gamma\}$ characterizing the environment. Given any environment E, the parameters ε_γ can be estimated from samples of the payoff function. Moreover, the larger the m associated with a given γ, the more samples required to reach a given level of confidence in the estimate of ε_γ. Hence, at any time, an adaptive strategy can use only those controls $\{\theta_\gamma\}$ for which it has a reasonable estimate of the corresponding ε_γ. In this way a possible *regressus ad infinitum* in the control hierarchy is forestalled.

If $\zeta(t)$ in Eq. (4.2) is selected to approach r_{*E} from below at an ever-decreasing rate, Eq. (4.1) can be differentiated to yield a condition on $r_{\tau,E}$ assuring that the associated adaptive strategy is strictly (E) b-optimal:

$$\dot{r}_{\tau,E} \geq -r_{*E}\dot{\zeta}(t). \tag{4.4}$$

Thus combining Eqs. (4.4) and (4.3), we obtain a condition on the parameters $\{\theta_\gamma\}$ and $\{\varepsilon_\gamma\}$, using the bounding function $\zeta(t)$, to assure strict b-optimality:

$$\sum_{x \in X} \dot{P}_x \lambda_x + \sum_{\gamma \in \Gamma} P_\gamma \varepsilon_\gamma \frac{d \ln \theta_\gamma}{dt} \geq -r_{*E}\dot{\zeta}(t). \tag{4.5}$$

Since $\zeta(t)$ can be chosen from any of a wide range of functions, inequality

(4.5) often proves relatively easy to use—given a recursive definition of τ (and hence of the O_γ) and a recursive definition of the elements of $\mathscr{E}$, there often exist $\zeta(t)$ for which the dominance of the left side of (4.5) over the right side can readily be established. One example of a set of bounding functions, which satisfy several auxiliary conditions assuring that some realizable strategies τ can dominate the functions over some environments, is

$$\dot{\zeta}(t) = \frac{2k_1}{r_{*E}} \left[\frac{1}{(t+k_2)^3} - \frac{3(r_{*E}-1)(t+k_2)^2 - k_1}{[(r_{*E}-1)(t+k_2)^3 - k_1(t+k_2)]^2} \right] \tag{4.6}$$

where k_1 and k_2 are parameters and

$$\lim_{t\to\infty} \prod [1-\zeta(t)] = \operatorname{glb}_t \prod [1-\zeta(t)] = \exp(-k_1/k_2). \tag{4.7}$$

Using (4.6) for the right-hand side of (4.5), and using Eq. (3.4) for $\dot{P}_x$, constants k_1 and k_2 can be found such that (4.5) is satisfied given any class of linear environments $\mathscr{E}$ [$\varepsilon_\gamma = 0$ for all γ, hence the second term on the left of (4.5) vanishes]. That is,

$$\sum_{x\in X} \dot{P}_x \mu_i(\sigma_j) > -r_{*E}\dot{\zeta}(t), \tag{4.8}$$

since the left side involves exponential functions of t whereas the right side involves primarily a function of t^3. It follows at once that the adaptive strategy of Sec. III, when confronted with a class $\mathscr{E}$ of linear environments, is strictly ($\mathscr{E}$) b-optimal, $b = \exp[-k_1(\mathscr{E})/k_2(\mathscr{E})]$. This corresponds to a common observation, proven in a wide variety of specific instances, that independent sampling (independent maximization along each dimension) converges and is efficient for linear environments.

Consider now a simple nonlinear environment E with payoff function

$$\mu(\sigma_{h_1}\sigma_{h_2}\cdots\sigma_{h_m}) = \sum_i \mu_i(\sigma_{h_i}) + \delta(\sigma_{h_1})\,\delta(\sigma_{h_m})$$

where

$$\begin{aligned} \delta(\sigma_h) &= 1 && \text{if } h = 1 \\ &= -1 && \text{if } h = k \\ &= 0 && \text{otherwise.} \end{aligned}$$

It is readily established that, regardless of the value of ε, the least-mean-square estimators $\{\lambda_x\}$ satisfy the equations

$$\lambda_x = \mu_i(\sigma_j) \qquad \text{for all} \quad x = (i, j).$$

Let $\mu_1(\sigma_2) > \mu_1(\sigma_j)$ for all $j \neq 2$. If $\varepsilon = 0$, a string with σ_2 at the first position will maximize payoff. However, if ε is large enough, a string with σ_1 at the first position will maximize payoff. Let E be of this latter type. Then a strategy employing only least-mean-square estimators, such as the strategy of Sec. III, when confronted with E will converge to a "false peak" (payoff less than r_{*E}). Improvement is still bounded below by an exponential function, but too small an asymptote is approached.

Note that with E as given, some second-order ε_γ will be greater than zero. For example, $\varepsilon_{(1,1)(2,1)}$ will be greater than zero by an amount determined by ε and the $\{P_x\}$. Once $\varepsilon_{(1,1)(2,1)}$ is detected it can be utilized by causing $P_{(1,1)(2,1)}(t)$ to take values progressively larger than $P_{(1,1)}(t) \cdot P_{(2,1)}(t)$. Since, by definition,

$$0_{(1,1)(2,1)}(t) = \frac{P_{(1,1)(2,1)}(t)}{P_{(1,1)}(t) \cdot P_{(2,1)}(t)}$$

this causes $0_{(1,1)(2,1)}(t)$ to increase and hence the term

$$P_{(1,1)(2,1)}\varepsilon_{(1,1)(2,1)} \frac{d \ln 0_{(1,1)(2,1)}(t)}{dt}$$

constitutes a positive increment on the left side of (4.5), speeding convergence.

An adaptive strategy exploiting the technique just exemplified can achieve strict b-optimality over nonlinear environments characterized by quite complicated sets $\{\varepsilon_\gamma\}$. In fact a strategy using analogs of crossover, recombination, inversions, intrachromosomal gene duplication, etc. can, for quite a wide range of sets of $\varepsilon_\gamma \neq 0$, sample the ε_γ and modify the 0_γ so that the asymptote approached exponentially is revised upward rapidly enough to preserve an overall exponential approach to r_{*E}, albeit a slower one than in (4.8). For example, the crossover rate at any point can be made subject to selection on the basis of sampled ε_γ. A low crossover rate between two substrings, indexed by γ_1 and γ_2, increases their dependence and hence increases factors like $0_\gamma = P_\gamma / P_{\gamma_1} P_{\gamma_2}$ where γ is the index of the concatenation of the two substrings. (In effect this gives the adaptive strategy a means of preserving valuable substrings, i.e. subroutines or nonprimitive components.) Thus, mechanisms like crossover can provide a sequence of increments of the form

$$P_\gamma \varepsilon_\gamma \frac{d \ln 0_\gamma}{dt}.$$

These increments increase the left side of (4.5) rapidly enough to assure that $\zeta(t)$ of (4.6), with an appropriate choice of k_1 and k_2, serves as a lower bound for all t. Strict b-optimality results.

A strict ($\mathscr{E}$) b-optimal strategy is *compared*, in effect, to a "gnostic strategy" which, confronted with any $E \in \mathscr{E}$, immediately selects the best device in $\mathscr{A}$

(highest payoff rate). This "selection" is made in the absence of any information about which element of $\mathscr{E}$ is present—hence, a "gnostic strategy" is not physically realizable unless a single device in $\mathscr{A}$ is best for all $E \in \mathscr{E}$. To be physically realizable a strategy must at least receive enough information from E to distinguish E from other elements of $\mathscr{E}$ which require other devices from $\mathscr{A}$ to achieve the best payoff rate. Acquisition of this information requires time, the duration depending upon the type of information transmitted from E and the manner of processing it. $\mathscr{E}$ may be quite "hostile" in this respect. For example it may contain several elements each having a different state with exceptionally high payoff but relatively isolated under initial conditions—a "spectral line" or isolated "pinnacle." To locate this state (to determine which $E \in \mathscr{E}$ is present) is, a priori, equally difficult for all realizable strategies, even though the "gnostic strategy" would exploit the "pinnacle" from the outset. Thus a strategy of strict b-optimality over such $\mathscr{E}$ may fail and yet be $(T, \mathscr{E})$ b-optimal with respect to realizable strategies T. Stated another way: If $\tau_0 \in T$ is strictly b-optimal over some class $\mathscr{E}$, and T consists only of realizable strategies, then τ_0 will be b-optimal relative to T over a much larger class $\mathscr{E}_T \supset \mathscr{E}$. This provides a natural way of exploiting a proof of strict ($\mathscr{E}$) b-optimality, for a strategy τ_0, to determine the enlarged class of environments $\mathscr{E}_T$ for which τ_0 is $(\mathscr{E}_T, T)$ b-optimal.

ACKNOWLEDGMENT

A major portion of the research reported here was supported by the National Institutes of Health through grant GM12236.

REFERENCES

[1] Ashby, W. R., "Design for a Brain," 2nd ed. Wiley, New York, 1960.
[2] Bellman, R., "Adaptive Control Processes: A Guided Tour." Princeton Univ. Press, Princeton, New Jersey, 1961.
[3] Bledsoe, W. W., and Browning, I., Pattern recognition and reading by machine. *Proc. Eastern Joint Computer Conf., Boston, 1959* pp. 225–232 (1959).
[4] Fisher, R. A., "The Genetical Theory of Natural Selection." Dover, New York, 1958.
[5] Fraser, A. S., Simulation of genetic systems. *J. Theoret. Biol.* 2, 329–346 (1962).
[6] Friedberg, R. M., A learning machine, Pt. I. *IBM J. Res. Develop.* 2, 2–13 (1958).
[7] Gelernter, H. L., and Rochester, N., Intelligent behavior in problem-solving machines. *IBM J. Res. Develop.* 2, 336–345 (1958).
[8] Goldberg, A. L., and Wittes, R. E., Genetic code: Aspects of organization. *Science* 153, 420–424 (1966).
[9] Grant, V., "The Origin of Adaptations." Columbia Univ. Press, New York, 1963.
[10] Kimura, M., On the change of population fitness by natural selection. *Heredity* 12, 145–167 (1958).
[11] Mayr, E., "Animal Species and Evolution." Harvard Univ. Press, Cambridge, Massachusetts, 1963.

[12] Milner, P. M., The cell assembly, Mark II. *Psychol. Rev.* 64, 242–252 (1957).
[13] Minsky, M., Steps toward artificial intelligence. *Proc. IRE* 49, 8–30 (1961).
[14] Moran, P. A. P., "The Statistical Process of Evolutionary Theory." Oxford Univ. Press, London and New York, 1962.
[15] von Neumann, J., *in* "Theory of Self-Reproducing Automata" (A. W. Burks, ed.). Univ. of Illinois Press, Urbana, Illinois, 1966.
[16] von Neumann, J., and Morgenstern, O., "Theory of Games and Economic Behavior." Princeton Univ. Press, Princeton, New Jersey, 1947.
[17] Newell, A., Shaw, J. C., and Simon, H. A., A variety of intelligent learning in a general problem solver, *in* "Self-Organizing Systems." Pergamon Press, New York, 1960.
[18] Rochester, N., Holland, J. H., Haibt, L. H., and Duda, W. H., Tests on a cell assembly theory of the action of the brain, using a large digital computer. *IRE Trans. Inform. Theory* IT-2, 80–93 (1956).
[19] Samuel, A. L., Some studies in machine learning using the game of checkers. *IBM J. Res. Develop.* 3, 210–229 (1959).
[20] Selfridge, O. G., Pandemonium, a paradigm for learning. *Mechanization of Thought Processes, Natl. Phys. Lab. Symp. No. 10.* Her Majesty's Stationery Office, London, 1959.
[21] Steck, G. P., Stochastic model for the Browning–Bledsoe pattern recognition scheme. *IRE Trans. Electron. Computers* EC-11, 274–282 (1962).
[22] Tax, S. (ed.), "The Evolution of Life." Univ. of Chicago Press, Chicago, Illinois, 1960.
[23] Turing, A. M., The chemical basis of morphogenesis. *Phil. Trans. Roy. Soc. London* B237, 37–77 (1952).

THE UNIVERSITY OF MICHIGAN
COLLEGE OF LITERATURE, SCIENCE, AND THE ARTS
Computer and Communication Sciences Department

Technical Report

HIERARCHICAL DESCRIPTIONS, UNIVERSAL SPACES AND ADAPTIVE SYSTEMS

John H. Holland

ORA Projects 01252 and 08226

supported by:

Department of Health, Education, and Welfare
National Institutes of Health
Grant No. GM-12236-03
Bethesda, Maryland

and

U.S. Army Research Office (Durham)
Grant No. DA-31-124-ARO-D-483
Durham, North Carolina

administered through:

OFFICE OF RESEARCH ADMINISTRATION ANN ARBOR

August 1968

HIERARCHICAL DESCRIPTIONS, UNIVERSAL SPACES AND ADAPTIVE SYSTEMS

John H. Holland

The power of an adaptive system depends critically upon its ability to exploit common factors in successful techniques. If the system has meager means for analyzing elements of its repertory, this ability will be sharply curtailed, no matter how extensive the repertory. Contrariwise, if the system has a great many different ways of describing (or representing) the same device, i.e., if it has a rich variety of ways to decompose elements of its repertory, chances of detecting common factors are greatly enhanced. Each time a device is tried, information accrues about components of each of the potential decompositions. Thus, the richer the variety of decompositions, the higher the effective sampling rate. Of course, to exploit this information about components, the adaptive system must use it to infer the performance of untried devices. And these inferences must, in turn, be used to plan which devices should be generated and tried next. At each stage, the flexibility and success of the process depends upon the flexibility and richness of the system's analysis and synthesis procedures--qualities ultimately depending upon the definitions of structure employed by the system.

In attempting to supply a rich set of representations (with attendant analysis and synthesis procedures), it helps to look at procedures actually exploited by successful adaptive plans. Among the most important are:

(1) Substitution--components common to several highly-rated devices are substituted in other related devices.

(2) Abstraction--by a process of abstraction, highly-rated devices are used to provide schemata (patterns for substitution) for the development of related devices.

(3) Refinement--new "cues" for reacting to the environment are provided by refining the input alphabet or time-scale, adjusting internal processing accordingly.

(4) Modeling--the environment is approximated by some part of the internal structure with the intention of checking predictions of this model against observed outcomes and modifying it accordingly.

(5) Change of Representation--new primitives and operations are introduced so that problems presented by the environment are more easily modeled and related to previous models, outcomes, or stored information.

(6) Metacontrol--rules for employing the preceding techniques are implemented in the device and they in turn are subject to the same techniques; additional levels are added as required.

By searching for structural traits which will give these techniques broad scope, we obtain suggestions for requirements on the structural formalism. These requirements will be briefly described here, and then each in turn will be discussed at length.

1. Hierarchical Description. Substitution and abstraction can be greatly facilitated if the devices used by the adaptive plan have many alternative descriptions in terms of "block diagram" hierarchies. A great variety of schemata can then be formed from a single device by simply deleting the contents of one or more blocks in different ones of its hierarchical descriptions. Any device which satisfies the input-output interface of such an "empty" block becomes a candidate for substitution therein.

Refinement proceeds most easily if alphabets and time-scales of blocks in the hierarchical description can be changed without requiring overall structural reorganization. The minimal constraints possible are those imposed by interfaces with other blocks and by identifications with parts of higher-level blocks.

2. Self-Applicability. The operations for connecting blocks should themselves be defined by exactly the same hierarchical descriptions as the objects to which they are applied (permitting new operations to be introduced as needed). Operations should apply equally directly to blocks at any level in the hierarchical descriptions (permitting any block to be treated as primitive). Taken together these provisions permit the adaptive plan to add new levels of control as required.

3. Incorporation of Models. Primitives and operations should be such that models of the environment can easily be implemented, used, and altered by the overall adaptive plan. There should be a clear means of designating the control exercised by active portions (cf. active subroutines) of the model. There should also be natural structural provisions for making and recording predictions based upon the model. These provisions permit the adaptive plan to check predictions against outcome and make corresponding modifications of responsible parts of the model (a technique used to great advantage by Samuel).

"Some Studies of Machine Learning, Using the Game of Checkers", and "Recent Progress"

Beyond these requirements, there is a further structural consideration which greatly aids study of the interaction of adaptive plan and environment. Because of von Neumann's unfinished work (See Burks' 1967 edited version)

<u>Theory of Self-Reproducing Automata</u>, Edited and completed by A. W. Burks

---

we know that it is possible to represent any mechanistically or computationally definable adaptive plan/environment combination in a uniform format--a format akin to the "space" of physics with its regular geometry and uniform laws. Such a "space" permits a uniform and rich characterization of both environments and plans. Of greater importance, in this "space" all actions are of the same kind, whether of the plan on itself, or within the environment, or between plan and environment. This and additional advantages will be developed in the discussion of self-applicatility (Section 2).

1. Hierarchical Descriptions

An approach to hierarchical descriptions first requires a notion of the devices or systems to be so described. Discussion here will be limited to systems constructed from finite-state components. That is, admissible components will be those for which relevant behavior can be determined in terms of a finite number of states. The formal counterpart of such a component is the finite automaton defined here by the quintuple

$$a = \langle I, S, 0, f, u \rangle$$

where

$$I = \prod_{i=1}^{m} I_i, \text{ where each } I_i \text{ is a finite set}$$

$$S \text{ is a finite set}$$

$$0 = \prod_{j=1}^{n} 0_j, \text{ where each } 0_j \text{ is a finite set}$$

$$f: \quad I \times S \rightarrow S$$

$$u: \quad I \times X \rightarrow 0.$$

Under the intended interpretation, I_i is the set of signals possible on the ith input (line) to the component, 0_j is the set of signals possible on the jth output, and S is the component's set of internal states. If $I(t) = [I_1(t),\ldots,I_m(t)]$ designates the (ordered) set of signals on the inputs at time t, and S(t) designates the internal state at time t, then the set of output signals at time t, $O(t) = [0_1(t),\ldots,0_n(t)]$, is given by

$$O(t) = u[I(t), S(t)].$$

-5-

Similarly, the next state of the automaton is given by

$$S(t+1) = f[I(t), S(t)].$$

A <u>complete input history</u> of the finite automaton, $[I(0), I(1), \ldots, I(t), \ldots]$, is an element of the set I^N of all infinite sequences over I. If we know the initial state S(0) and a complete input history, it follows from the definition of f and u that we can determine the complete sequence of internal states and the complete sequence of output signals, elements of S^N and O^N respectively. That is, f and u can be extended to yield the functions

$$F: \quad I^N \times S \to S^N$$

and

$$U: \quad I^N \times S \to O^N$$

where, formally, $N = \{0,1,2,\ldots\}$ and $I^N = \{\underline{I} \ni \underline{I}: \ N \to I\}$, etc.

Our interest will be centered on devices which can be constructed by various means from copies of some intially chosen finite set of automata, $G = \{g_1, \ldots, g_k\}$, which will be called a <u>set of generators</u> (or primitives). Let $\{a_\alpha \ni \alpha \in A\}$ be a finite or countably infinite set of copies of elements drawn from G and indexed by the ordered set A. The elements of the quintuple corresponding to a_α will be designated I_α, S_α, etc.

We obtain the most easily understood of the systems constructed from the (entire, possibly infinite) collection $\{a_\alpha\}$ if the collection is treated as a single automaton composed of the noninteracting automata a_α. In this automaton the state history or output history of any given component is independent of the histories of all other components. This "free product", or <u>unrestricted composition</u>, $\underline{A}$, on $\{a_\alpha\}$ is defined by the quintuple $<I_A, S_A, O_A, f_A, u_A>$. Here

-6-

$$S_A = \prod_{\alpha\varepsilon A} S_\alpha .$$

A second (equivalent) definition of S_A as a collection of functions

$$S_A = \{s: A \rightarrow \bigcup_\alpha S_\alpha \ni \text{for all } \alpha\varepsilon A,\ s(\alpha)\ \varepsilon S_\alpha\}$$

provides a convenient notational device; for instance, if s is the state of the automaton $\underline{A}$, then the state of the set of components indexed by $A' \subset A$ is simply the restriction of s to A', $s|A'$. Similarly

$$I_A = \prod_{\alpha\varepsilon A} \prod_{i=1}^{m_\alpha} I_{\alpha,i} = \{i: X \rightarrow \bigcup_x I_x \ni x\varepsilon X \text{ and } i(x)\varepsilon I_x\}$$

where $X = \{(\alpha,i) \ni \alpha\ \varepsilon\ A \text{ and } 1 \leq i \leq m_\alpha\}$,

and $$O_A = \prod_{\alpha\varepsilon A} \prod_{j=1}^{n_\alpha} O_{\alpha,j} = \{o: Y \rightarrow \bigcup_y O_y \ni y\varepsilon Y \text{ and } o(y)\varepsilon O_y\}$$

where $$Y = \{(\alpha,j) \ni \alpha\varepsilon A \text{ and } 1 \leq j \leq n_\alpha\}.$$

$f_A: I_A \times S_A \rightarrow S_A$ satisfies the requirement

$$f_A(i,s)\ (\alpha) = s'(\alpha) = f_\alpha[i(\alpha),\ s(\alpha)]$$

where

$$i\varepsilon I_A,\ s \text{ and } s'\varepsilon S_A, \text{ and } i(\alpha) = [i(\alpha,1),\ldots,i(\alpha,m_\alpha)].$$

That is, f_A when applied to an input state i and an internal state s of $\underline{A}$ yields a new state s'; $i(\alpha)$, $s(\alpha)$, and $s'(\alpha)$ are the values of these states at the component a_α; since the components do not interact we can determine the value of $s'(\alpha)$ from $i(\alpha)$ and $s(\alpha)$ alone, using f_α. Similarly

$$u_A\colon I_A \times S_A \to O_A \text{ satisfies}$$

$$u_A(i,s)(\alpha) = o(\alpha) = u_\alpha[i(\alpha),\ s(\alpha)].$$

[FIGURE 1.]

Other systems or <u>compositions</u> over the collection $\{a_\alpha\}$ are obtained by imposing restrictions on the "free product" $\underline{A}$ --restrictions resulting from the identification (connection) of selected inputs with selected outputs. Under these restrictions the state or output history of a given component will, in general, depend upon the histories of other components. The identifications determining the restrictions on $\underline{A}$ can be specified by a function γ, the <u>composition function</u>, satisfying the conditions:

(i) $\gamma\colon Y' \to X'$ from $Y' \subset Y$, 1-to-1 onto $X' \subset X$;

(ii) $O_y \subset I_{\gamma(y)}$ for all $y \varepsilon Y'$.

Thus each $y \varepsilon Y'$ designates a particular output which is to be connected to a particular input $\gamma(y) \varepsilon X'$; (ii) specifies that the output signals be identifiable as signals on the input $\gamma(y)$ -- any encoding or decoding must be carried out by components via their functions f and u. Identification of an input with an output requires the histories of the two to be identical over all time:

$$U_y(\underline{I}_\alpha, s) = \underline{I}_{\gamma(y)},$$

where U_y is the function that gives the output history of $y = (\alpha, j)$ when supplied with the initial state s and the input history $\underline{I}_\alpha$ of the component α to which y belongs. Note that $\underline{I}_\alpha = (\underline{I}_{\alpha,1}, \ldots, \underline{I}_{\alpha,m_\alpha})$ and may in turn be constrained by identifications specified by γ. It may

$A = \{1, 2, 3\}$

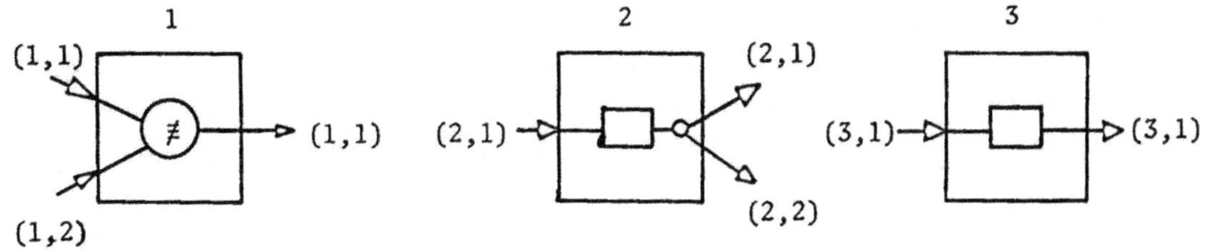

$S_1 = \{0\}$ $\quad$ $S_2 = \{0,1\}$ $\quad$ $S_3 = \{0,1\}$

$$S_A = \{0\} \times \{0,1\} \times \{0,1\}$$
$$= \{(0,0,0),\ (0,0,1),\ (0,1,0),\ (0,1,1)\}$$
$$= \{s \ni s:\ \{1,2,3\} \to \{0,1\}\} = \{s_0,\ s_1,\ s_2,\ s_3\}$$

where s_2, for example, is given by

α	$s_2(\alpha)$
1	0
2	1
3	0

I_A and O_A are treated similarly, with i_7, for example, given by

(α,j)	$i_7(\alpha,j)$
(1,1)	0
(1,2)	1
(2,1)	1
(3,1)	1

If $S_A(t) = s_2$ and $I_A(t) = i_7$ then

$$S_A(t+1)(1) = f_A(i_7,s_2)(1) = f_1\Big([i_7(1,1),i_7(1,2)],s_2(1)\Big)$$
$$= f_1([0,1],0) = S_1(t+1) = 0$$

and $S_A(t+1)(2) = f_A(i_7,s_2)(2) = f_2\Big([i_7(2,1)],S_2(2)\Big)$
$$= f_2([1],1) = S_2(t+1) = 1$$

Figure 1. An Unrestricted Composition.

-9-

even be that $\gamma(y) = (\alpha,i)$ in which case the output feeds back directly to one of the inputs of the component, yielding the equation

$$U_{(\alpha,j)}\left((\underline{I}_{\alpha,1},\ldots,\underline{I}_{\alpha,i},\ldots,\underline{I}_{\alpha,M_\alpha}),s\right) = \underline{I}_{\alpha,i} \,.$$

This amounts to a kind of fixed point requirement on the function U_y, a requirement which may or may not be satisfiable. (See Figure 2.)

[Figure 2.]

In the general case we get a set of simultaneous equations, one for each element of Y', which may or may not be consistent (i.e., satisfiable). Only if the equations are consistent will there exist a quintuple corresponding to the product automaton constrained by γ. The consistency of γ will be assured if it satisfies an appropriate "local effectiveness" condition in addition to conditions (i) and (ii). This condition can be developed along the following lines:

A sequence of output indices

$$y_1 = (\alpha_1,j_1),\ y_2 = (\alpha_2,j_2),\ \ldots,\ y_k = (\alpha_k,j_k)$$

will be called <u>connected</u>, and of length k, if

$$(1 \le h < k)(\exists b_h)[\gamma(y_h) = (\alpha_{h+1},b_h)].$$

γ will be called consistent with respect to $s \varepsilon S_{\underline{A}}$ if, for each $\alpha \varepsilon A$, there is an integer $\ell_{s,\alpha}$ such that every connected set of output indices of length $\ell_{s,\alpha}$ which ends at α contains some $y_h = (\delta,j)$ for which $u_{\delta,j}(i,s)$ depends only on unconstrained inputs, i.e.,

The logical net

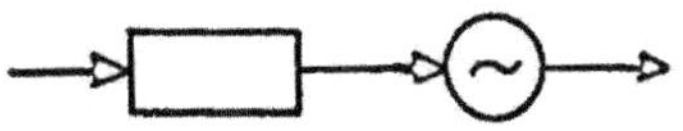

has the fixed point $\underline{I}$ = 101010... since U($\underline{I}$,0) = 101010... .

The logical net

has no fixed point since for $\underline{I} = \delta_0\ \delta_1\ \delta_2\ \ldots$, $U(\underline{I},0) = \bar{\delta}_0\ \bar{\delta}_1\ \bar{\delta}_2\ \ldots$

where $\bar{\delta}_j = 1 - \delta_j$.

Figure 2. Examples of Satisfiable and Unsatisfiable Cases of the Fixed Point Requirement.

$$(i_1, i_2 \varepsilon I_A)[i_1|(X-X') \Rightarrow u_{\delta,j}(i_1(\delta), s(\delta)) =$$

$$u_{\delta,j}(i_2(\delta), s(\delta)].$$

The inputs of the product automaton indexed by the set X - X' are unconstrained by γ and hence may be arbitrarily specified at each time t. For this reason, the set

$$I_{\gamma,A} = \{i|X-X') \ni i \varepsilon I_A\}$$

plays the role of the set of input states of the constrained automaton. Similarly the elements of

$$O_{\gamma,A} = \{o|(Y-Y') \ni o \varepsilon O_A\}$$

play the rôle of output states.

Modifying an algorithm of Burks and Wright,

"Theory of Logical Nets"

---

one can prove

Lemma 1.1. If γ is consistent with respect to s ε S_A, then there is a unique correspondence $\mu_s : I_{\gamma,A} I_A$ such that, for every i in the range of μ_s and all y ε Y', $u_y(i(\alpha), s) = i(\gamma(y))$. Thus, for any input state of the constrained automaton, when γ is consistent, there exists a unique assignment of states to the constrained inputs x ε X' such that the values $u_{\gamma^{-1}(x)}$, computed for the corresponding constrained outputs $\gamma^{-1}(x)$, are in fact identical as required.

γ will be called a locally effective composition function with respect to s, abbreviated LECF(s), if (i)γ is consistent with respect to s, and (ii) the consistency of γ with respect to any $s' \varepsilon S_A$ implies that γ is consistent with respect to $f_A(\mu_s,(i),s')$ for every $i \varepsilon I_{\gamma,A}$.

We obtain as a corollary to Lemma 1.1.

Corollary 1.1. If γ is LECF(s), the correspondence μ_s of Lemma 1.1 can be extended to a unique correspondence μ_s: $I^N_{\gamma,A} \to I^N_A$ such that for every $\underline{I}$ in the range of μ_s and all $y \varepsilon Y'$, $U_y(\underline{I}_\alpha, s) = \underline{I}_{\gamma(y)}$.

For each LECF(s) associated with the product automaton $\underline{A}$, the composition $\underline{A}$ is defined by the quintuple $<I_{\gamma,A}, S_{\gamma,A}, O_{\gamma,A}, f_{\gamma,A}, u_{\gamma,A}>$ where $I_{\gamma,A}$ and $O_{\gamma,A}$ are as above

$S_{\gamma,A} = \{s \varepsilon S_A \ni \gamma$ is consistent with respect to $s\}$

$f_{\gamma,A}(i,s) = f_A(\mu_s(i),s)$ for $i \varepsilon I_{\gamma,A}$ and $s \varepsilon S_{\gamma,A}$

$u_{\gamma,A}(i,s) = u_A(\mu_s(i),s)$.

When particular compositions are discussed, the subscript A will be dropped where no confusion can arise. The set of compositions, $\{\underline{A}_\gamma \ni \underline{A}$ is an unrestricted composition and γ is an associated LECF(s) composition function$\}$, will be the set of devices or systems for which we will develop hierarchical descriptions. The set includes representatives of many well-known devices including logical nets, Turing machines with multiple heads and multi-dimensional tapes, and variants of von Neumann's cellular automaton. In the case of devices such as Turing machines and cellular automata, the index set A is countably infinite. When A is infinite the local effectiveness condition assures a critical property--one which permits investigation of finite parts of the overall system with knowledge only of the states of components in a finite "neighborhood" thereof:

Lemma 1.2. Given any $\underline{A}_\gamma$, $\alpha \varepsilon A$, and $t \varepsilon N$, $S_\gamma(t)|\alpha$ can be calculated from

$S_\gamma(0)|B(\alpha,t)$ and $I_\gamma(0)|B(\alpha,t)$, ..., $I_\gamma(t)|B(\alpha,t)$, where B is a computable function and $B(\alpha,t)$ is a finite subset of A containing α.

We now have a formal definition of the structures which will be candidates for hierarchical description. A given finite composition will have hierarchical descriptions much like a cross-referenced set of increasingly-detailed block diagrams--the kind of diagrams used to describe almost any very complex organization. The highest-level block diagram divides the system into several large, interacting parts. The guiding principle in making this division is a concern that the parts exhibit a "natural" functional coherence in terms of intended use or overall function. It is apparent that different intended uses or different views of the overall function can lead to different highest-level diagrams of the same device. We can thus expect each device to have many distinct hierarchical descriptions. Once the highest-level diagram is set, each of its parts is treated in turn as a system to be further subdivided. The process of subdivision is repeated, generating successive levels of refinement, until parts are reached simple enough not to repay further subdivision. By way of example, a hierarchy of 11 levels in which each block is divided into 10 lower-level blocks would contain 10^{10} lowest-level blocks (about the number of neurons in the human brain). If each of the lowest-level blocks contained a single two-state device, the overall device would have $2^{10^{10}}$ states. Even for a device with as many as 10^{10} components, one need only make a selection at each of 10 levels to uniquely locate any given component. And, assuming a relevant functional division, much will be learned of the effect of that component by observing the use or function of the blocks involved. In contrast, an explicit description of the device in terms of a state transition diagram is not even a possibility; the number of states involved vastly exceeds the estimated number of atoms in our galaxy. Moreover, presentation in terms of states can be misleading when we study various operations important to adaptation. A state reduction from 1,000,000 to 500,000 states looks impressive, but it may result from the elimination of a single two-state device from a connected set of 20-two-state devices. It may seem impossible to begin with a set of 10^{1000} states and give them any very significant organization and yet, in terms of components, this is achieved almost routinely. Even a small digital computer has a much larger number of states. For devices of this complexity, hierarchical descriptions offer almost the only avenue to detailed understanding.

It will be convenient to define each hierarchical structure formally via a directed tree. Following our previous discussion, each vertex in the tree corresponds to a block in the hierarchical structure. Two vertices, α_1 and α_2, are connected by a directed edge from α_1 to α_2 just in case the block corresponding to α_2 is a part of the block corresponding to α_1. The tree has a root (distinguished vertex) with only outgoing edges, representing the whole device. It also has a set of terminal vertices, representing the blocks at the level of finest detail. Otherwise, vertices have one incoming edge and one or more outgoing edges. The vertices connected to a vertex by its outgoing edges will be called its <u>successor set</u>; each successor set will be ordered. By listing in order the ordinals of vertices in the unique path from the root to any given vertex, we obtain a unique index for each vertex. Thus, the second vertex in the successor set of the root will have the index 1.2, and the first vertex of the successor set of 1.2 will have the index 1.2.1, etc.

We will now examine a simple set of hierarchical descriptions. This prototype, though simple, adequately illustrates the use of the tree format and will give a concrete background for later discussion. Much more sophisticated descriptions can readily be constructed on the same pattern. Still, the prototype descriptions exhibit the essential property of progressive refinement of the description as levels are added to the hierarchy. That is, as levels are added, the fineness of the time-scale generally increases and constraints are added to the input and output alphabets and to the transition function. From the behavioral viewpoint, this means that an adaptive system employing these descriptions can refine its responses by simply adding levels to the hierarchy describing its current plan. Moreover, in the formation of schemata, the plan can control the range of substitution instances by controlling the level at which blocks are deleted.

Each block in the hierarchical structure will be treated as a composition, although elements of its associated quintuple may be specified only when other parts of the hierarchical structure are specified. The parts of a block may thus be interconnected (by a LECF composition function) leaving only a subset of their inputs (outputs) free. (Recall that the successor set of the vertex associated with a given block designates the parts of that block). The free inputs (outputs) of the parts must be identified, in some order, with the free inputs (outputs) of the block. In effect, subsets of the free inputs (outputs) of the parts will be "cabled" and identified with some free input (output) of the block. Formally, a direct product will be formed of the alphabets of each "cabled" subset (in the order imposed) and this new alphabet will be associated with the designated line of the block. To present this information in precise format labels are attached to each vertex of the directed tree as follows:

Let α be the index of an arbitrary vertex and let $\alpha = \alpha'.k$ so that the vertex is the kth vertex in the successor set of α'. The block associated with α will have m_α inputs and n_α outputs; accordingly the vertex α will be labelled by two vectors,

$$v_\alpha = (v_{\alpha,1},\ldots,v_{\alpha,m_\alpha})$$

$$w_\alpha = (w_{\alpha,1},\ldots,w_{\alpha,n_\alpha})$$

where

$v_{\alpha,j} = (\ell,j_1)$ if $\gamma^{-1}(\alpha,j) = (\alpha'.\ell,\ j_1)$, that is if input j of α is connected to output j_1 of block $\alpha'.\ell$ which is also a part of block α'

$= (j_o)_h$ if $I_{\alpha,j} = \text{proj}_h\ I_{\alpha',j_o}$, that is if input j of block α is component h of a "cable" identified with input j_o of block α'.

$w_{\alpha,j} = (\ell,j_1)$ if $\gamma(\alpha,j) = (\alpha'.\ell,j_1)$

$= (j_o)_h$ if $O_{\alpha,j} = \text{proj}_h\ O_{\alpha',j_o}$.

See Figure 3.

[Figure 3.]

To allow arbitrary relations between the time-scales (the internal clocks) of different blocks--for example, to allow one block to execute several operations on receipt of each signal from another--we must extend this formalism. So that there will be no absolute limit to the refinement of time-scales, it is preferable to specify block time-scales in relation to one another, rather than in relation to some absolute. To accomplish this, let us adopt the convention that the elements of any input alphabet

The logical net

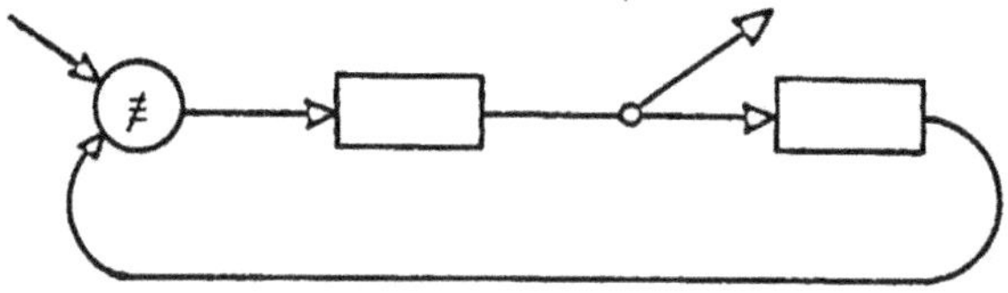

has as one of its hierarchical descriptions the diagram

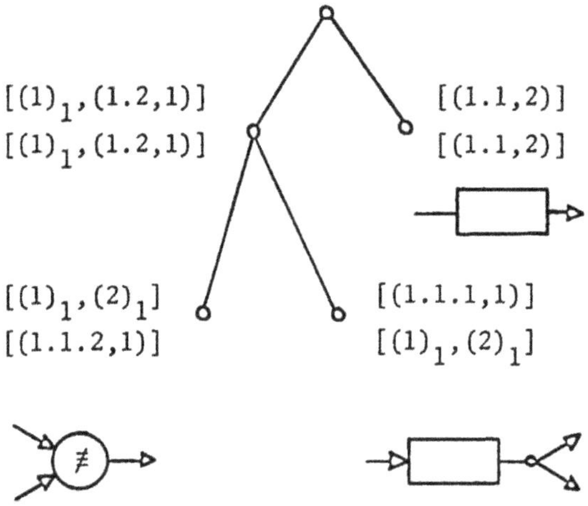

where inputs and outputs are ordered from top to bottom (in each block) and and vertices are ordered from left to right (at each level).
(Note that cabling and rates are not exemplified).
Deletion of the two lowest level vertices in the hierarchical description yields a schema which can be represented as the incompletely specified logical net.

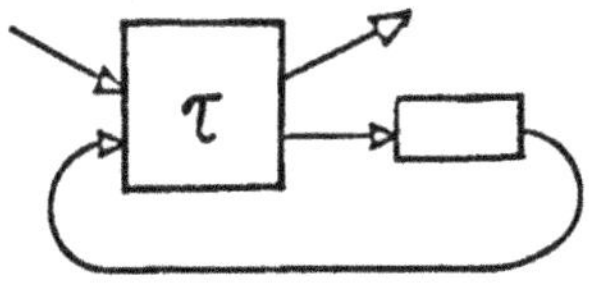

Figure 3. Example of a Hierarchical Description.

$I_{\alpha,j}$ may in fact be a set of strings of some fixed length $k_{\alpha,j}$. Treat output alphabets similarly. Then relax the requirement $O_{\alpha,j} \subseteq I_{\gamma(\alpha,j)}$ to the requirement that, for some r, $[O_{\alpha,j}]^r \subseteq I_{\gamma(\alpha,j)}$ or else $O_{\alpha,j} \subseteq [I_{\gamma(\alpha,j)}]^r$; that is, the elements of $I_{\gamma(\alpha,j)}$ are strings of length $rk_{\alpha,j}$ formed by concatenation of strings in $O_{\alpha,j}$, or vice-versa. With this convention, r time-steps in α will be required to generate a single input symbol for the input line $\gamma(\alpha,j)$ or alternatively, one time-step in α will generate r input symbols for $\gamma(\alpha,j)$. In compositions, relative rate will be indicated by labelling each output (α,j) with the value $\delta(\alpha,j)$, where $\delta(\alpha,j) = r$ when $[O_{\alpha,j}]^r \subseteq I_{\gamma(\alpha,j)}$ and, by convention $\delta(\alpha,j) = \frac{1}{r}$ when $O_{\alpha,j} \subseteq [I_{\gamma(\alpha,j)}]^r$. These values will be indicated in the hierarchical description by augmenting $v_{\alpha,j}$ and $w_{\alpha,j}$; thus

$v_{\alpha,j} = \left((\ell,j_1),\frac{1}{r}\right)$ if $\gamma^{-1}(\alpha,j) = (\alpha'.\ell,j_1)$ and output $\gamma^{-1}(\alpha,j)$ has $\frac{1}{r}$ as its rate indicator, that is $O_{\gamma^{-1}(\alpha,j)} \subseteq [I_{\alpha,j}]^r$

All other cases are treated similarly.

There are rate assignments which make inconsistent requirements on the time-scales of blocks in the hierarchy; however, there exists a convenient necessary and sufficient condition for assuring consistency. It can be developed as follows:

Let $\delta_1, \delta_2, \ldots, \delta_k$ be the rate indicators associated with an arbitrary connected set of output indices $y_1 = (\alpha_1,j_1), \ldots, y_k = (\alpha_k,j_k)$. With each connected sequence we can associate the product $\pi_h \delta_h$. Consistency of time-scales requires:

(i) Any two connected sequences which start at the same block and end at the same block must have the same value for their associated products.

(ii) Any cycle (a connected sequence which begins and ends at the same block) must have the value 1 for its associated product.

These conditions must be satisfied even for connected sequences which are "hidden" (implicit) in "cables".

Consider now a directed tree with labels v_α and w_α, as just defined, assigned to its vertices. The result is a kind of organizational skeleton which can be fleshed out by assigning specific procedures to the terminal vertices. In the present context this is simply accomplished by assigning to a terminal vertex β any composition with the number of inputs and outputs required by v_β and w_β. If one or more terminal vertices in the directed tree are left with no composition assigned, the result will be called a description schema. If all terminal vertices have compositions assigned, then we must determine whether the result in fact describes a composition. To do this we must first determine a composition function from the hierarchical description. This can be done recursively:

Assume that the composition functions $\gamma_{\alpha.1},\ldots,\gamma_{\alpha.b_\alpha}$ have been determined for parts $\alpha.1,\ldots,\alpha.b_\alpha$ of block α. Let (β_1,k_1) be an arbitrary output line identified as component h_1 of line $(\alpha.\ell_1,j_1)$ and let (β_2,k_2) be an arbitrary input line identified as component h_2 of line $(\alpha.\ell_2,j_2)$. Then the composition function for block α is determined bv the following condition on pairs of input and output lines. $\gamma_\alpha(\beta_1,k_1) = (\beta_2,k_2)$ if and only if

(i) $\gamma_{\alpha.\ell_1}(\beta_1,k_1) = (\beta_2,k_2)$, i.e., the lines are already connected in part $\alpha.\ell_1$;

(ii) $v_{\alpha.\ell_1} = [(\alpha.\ell_2,j_2),r]$ and $h_1 = h_2$, i.e., the "cables" containing the lines are connected at this level and the lines occupy corresponding positions in the two cables.

(All lines (β,k) not belonging to the domain or range of γ_α will perforce be identified by some $v_{\alpha.\ell}$ or $w_{\alpha.\ell}$ as components of some line of α). The recursion is started at the terminal vertices by using the composition functions of their assigned compositions. The end-result of the recursion is the composition function γ_1 obtained for the root vertex at the end of the recursion.

Not every composition function derived in this way is LECF. Of course if the description is of an extant finite composition the derived composition function is that of the extant composition. But the description may be obtained by substitution of arbitrary compositions at the terminal vertices of a schema. Then whether or not the derived function is LECF depends upon the particular compositions assigned; if the function is LECF then a composition is indeed described, otherwise not. There are various sufficient conditions for assuring the LECF property. Perhaps the simplest is an addition to requirement (ii) for time-scale consistency: Each cycle must contain an element with delay (an element such that the output function, u, is a function of S only, i.e., u: $S \to O$). Later, when we discuss self-applicability, we will come upon a more important technique for assuring the LECF property. We will see that hierarchical descriptions can be "embedded" in certain compositions, of countably infinite index, generated by a single element--the counterparts of von Neumann's cellular automaton. Moreoever "construction" operations for modifying the descriptions and substituting in schema can be embedded in the same composition. When this is appropriately carried out we will find that no sequence of construction operations can yield a hierarchical description without the LECF property.

The class of hierarchical descriptions just presented, though simple, has several valuable features. Many of these features follow directly from the definition of the underlying compositions: All "interface" transformations, such as fan-in, fan-out, encoding, decoding, matching of cable components, and change of relative clock rate, are represented explicitly in the descriptions. Thus, fan-out--the sending of the same output signal over several lines--is accomplished by an explicit composition with one input line and several output lines; each output line y has the same associated function, u, which simply places each input signal, i, on the output line, i.e., $u_y(i,s) = u(i,s) = i$, independent of s. Encoding--for example, to transform the output signals of one device to form acceptable by another--is accomplished by a composition which takes as argument the signal to be encoded, o, and yields as value the encoded form, c(o), i.e., $u(o,s) = c(o)$. One device can be required to operate at k times the clock rate of another by using an "interface" composition which has an input alphabet of strings of length k over the output alphabet of the controller and an output alphabet identical to the input alphabet of the controlled device. Similar procedures hold for other interface requirements. The advantage of this explicitness becomes apparent when we wish to compare and manipulate descriptions or the underlying devices. Devices constructed independently can be made compatible, for use in some larger device, by addition of explicit interface devices. More will be said along this line presently.

2. Self-Applicability

In preparation for a discussion of self-applicability of hierarchical descriptions, let us preview the uses of description schemata.

First, note that elimination of any subtree or set of subtrees from a hierarchical description automatically produces a schema; each unassigned vertex so-produced indicates a place for substitution. The substitution instances are of two kinds--those which produce descriptions of compositions, and those which yield new schema. In both cases the substitution instances must satisfy the interface requirements imposed by the remainder of the hierarchical description. In the case of instances of the first kind, these requirements are restrictions on the number, ordering and alphabets of input and output lines, and restrictions on the relative clock rate. The admissible substitution instances of the first kind are the compositions, or henceforth their descriptions, which satisfy the restrictions at the unassigned vertex. Of course, an otherwise unsuitable composition may be modified by interface devices which transform inputs, outputs, and rates, sufficiently to meet the requirements. Substitution instances of the second kind are schemata which meet the interface requirements at the unassigned vertex, insofar as they apply. A substituted schema refines the original schema by the addition of new structural features. The refined schema thus admits only a subset of the substitution instances of the original schema.

The schema's particular significance for adaptive plans emerges when we assign a measure of performance to each device-environment combination. As a result, given any environment from the set of posibilities, each substitution instance of a schema is assigned a numerical value. If now there is a probability distribution over the substitution instances we can determine an expected value for the (instances of the) schema. The distribution corresponds to a kind of preference ordering on the instances--instances assigned a high probability, by the distribution, being favored. In the context of adaptation, then, this distribution plays a role quite analogous to simplicity orderings in inductive logic. That is, in the usual presentation of inductive logic, there exists an infinite set of incompatible hypotheses agreeing with any finite presentation of evidence. However, the evidence is taken to confirm the hypothesis which is simplest (according to the ordering) and that hypothesis is held (tentatively) until further evidence is gathered. It is as if the hypotheses were "tested" one by one in the order given, each being rejected in turn, until one is encountered which satisifies the evidence. In applied situations, the tentative hypothesis will be a source of predictions (consequences) which will, in turn, influence the actions taken to gather new evidence. In the case of schemata, the distribution over instances (stochastically) sets the order and intensity with which the various schema will be tested. If we think of the schemata as hypotheses about useful organizational principles, then such hypotheses will be tried and (provisionally) "confirmed" or "disconfirmed" in the (stochastic) order imposed by the distribution. That is, as instances of a schema are tested, an estimate can be made of its value; by making the distribution conditional on this estimate, the probability of particular future tests can be altered accordingly. The overall effect is that of "accepting" or "rejecting" the hypotheses corresponding to the schemata, future action being based on the high-valued ("accepted") schemata.

There will be much more to say about the generation of distributions both later in the discussion of self-application and still later in the discussion of modelling. For now the central point is the possibility of ranking schema according to estimates of their expected performance, onee a distribution is given.

Estimates of the expected performance of given schemata provide an adaptive plan with a natural basis for inferences about untried schemata and devices. These inferences can in turn guide the plan in its selection of devices (and schemata) to be tested next. (If the device selected for test is presented by a hierarchical description, the result of the test provides a sample point for each of the schemata which can be derived from the description). For instance, the plan may take several samples of refined versions of a schema having a high performance estimate relative to other schemata tested. Such refinements are obtained, as indicated earlier, by substituting new schemata at the terminations of the given schema. If the performance measure varies significantly over instances of the given schema, then restriction of samples to an appropriate subset of instances will yield a higher expected performance. That is, there is a reason for testing refinements of the given schema, searching for one which restricts the instances appropriately. The schemata constitute a (highly redundant) covering of the set of devices; the refinements of a given schema constitute a (highly redundant) covering of its instances. In these terms, the object of refinement is to search out elements of the cover progressively converging on devices of high performance. Refinement, however, is only one possible procedure. To give one other example, the plan may "step" from one element of the cover to another by first deleting part of a schema and then substituting a new part. There are many other modification procedures, each appropriate to particular sets of sample outcomes. The procedures to be applied in given situations must be decided, at a higher level, by the adaptive plan. Recalling our earlier discussion, this amounts to modification of the probability distribution over substitution instances.

How is the adaptive plan to exert the higher-level control implicit in the selection of procedures for selecting (modifying) devices? The introductory remarks suggested a direct route: Make the hierarchical descriptions self-applicable. Then certain of the devices employed by the plan can control the selection of other devices to be tested against the environment. Moreover, additional levels of control can be supplied as needed, in exactly the same way. But how do we enable devices, hierarchically described, to operate on descriptions and schema?

One way to assure self-applicability is to "embed" the adaptive plan, including the descriptions of the devices it employs, in a common "logical space". This possibility is a direct consequence of von Neumann's work on cellular automata. To examine it we need a definition of the "logical spaces" and of the process of "embedding" descriptions in these spaces. The required "logical spaces" can be defined as particular compositions of countably infinite index, generated by a single element. However, in order to distinguish the appropriate compositions, a prior definition of "embedding" is necessary. To specify the manner in which one device (the object) is embedded in (is simulated by) another (the image) is to supply a mapping whereby actions in the image device can be reinterpreted as actions of the object. Since our purpose is to embed detailed descriptions, we wish to preserve in the image not only overall behavior, but also local details of the action. That is, in the image, we wish to find a counterpart of every detail of the object's structure; and we wish to be able to determine from these counterparts everything that could be determined from the original parts. This can be accomplished for compositions as follows:

Assume composition $\underline{A}_\gamma$ is to be embedded in composition $\underline{D}_\xi$. Then the image of $\underline{A}_\gamma$ under the embedding is to be a subcomposition $\underline{B}_\zeta$ of $\underline{D}_\xi$, where subcomposition is defined by the requirements

(1) $\{b_\beta \ni \beta \in B\} \subset \{d_\delta \ni \delta \in D\}$

(2) $\zeta = \xi | Y_B$, where $Y_B = \{y \in Y' \ni Y'$ is the domain of ξ, $\text{proj}_1 y \in B$, and $\text{proj}_1\, \xi(y) \in B\}$. (When no confusion can arise the subcomposition will simply be denoted $\underline{D}_{\xi|B}$.)

Let A, X, Y and S_γ, I_γ, O_γ be the index and state sets, respectively, of the object composition $\underline{A}_\gamma$ and let B, X_1, Y_1 and S_ζ, I_ζ, O_ζ be the corresponding sets for the image subcomposition $\underline{B}_\zeta$. An <u>embedding</u> will be defined by a mapping ϕ from the sets A, X, Y and S_γ, I_γ, O_γ to subsets of B, X_1, Y_1 and the sets S_ζ, I_ζ, O_ζ, respectively, satisfying:

(1) Distinct {indices, states} map onto distinct {sets of indices, states}:

$$\alpha \neq \alpha_1 \Rightarrow \phi(\alpha) \cap \phi(\alpha_1) = \text{null set, etc.}$$

(2) Each index of a {free, bound} {input, output} of a given element maps onto corresponding indices of the image subset:

$$\text{proj}_1\, \phi(x) \subset \phi(\text{proj}_1 x)$$

$$x \in X\text{-}X' \Rightarrow \phi(x) \subset X_1 - X' \quad \text{and similarly for } y$$

(3) If $x = \gamma(y)$, then the same must hold for all indices in $\phi(x)$:

$$\phi[\gamma(y)] = \zeta[\phi(y)].$$

(4) If an element a_α of $\underline{A}_\gamma$ is assigned the same state by two states of the composition, then the same must hold true in the image:

$$s(\alpha) = s_1(\alpha) \Rightarrow \phi(s)|\phi(\alpha) = \phi(s_1)\ \phi(\alpha)$$

and similarly for i, $i_1 \in I$ and o, $o_1 \in O$.

(5) The transition and the output functions of the image of each element a_α of $\underline{A}_\gamma$ must faithfully represent the transition and output function, respectively, of a_α:

$$f_{\phi(\alpha)}[\phi(i)|\phi(\alpha),\phi(s)|\phi(\alpha)] = \phi f_\alpha[i(\alpha),s(\alpha)],$$

where $f_{\phi(\alpha)}$ is the transition function of the subcomposition indexed by $\phi(\alpha)$, and similarly for $u_{\phi(\alpha)}$.

One final requirement assures that the image subcomposition is immune to disturbances via signals over unassigned inputs in the image. $i \in [I_\zeta|\phi(x) - \phi(I_\gamma)|\phi(x)]$ (an "illegal" signal on an assigned line) is still permitted to cause aberrant behavior:

(6) When the state of $\underline{B}_\zeta$ is the image of a state of $\underline{A}_\gamma$, then f_ζ and u_ζ do not depend upon the states of unassigned inputs:

$$i|\phi(X-X') = i_1|\phi(X-X') \Rightarrow f_\zeta[i,\phi(s)] = f_\zeta[i_1,\phi(s)]$$

where $s \in S_\gamma$ and $i, i_1 \in I_\gamma$, and similarly for u_ζ.

See Figure 4.

[Figure 4.]

It should be noted that, under this definition, several connected elements in the image may be used to represent a single element in the object. We shall have use for a stricter notion: An <u>isomorphic embedding</u> of $\underline{A}_\gamma$ on $\underline{B}_\zeta$ is an embedding such that, for all elements in A, X, Y, S_γ, I_γ, 0_γ and B, X_1, Y_1, S_ζ, I_ζ, 0_ζ,

$$\alpha \simeq \phi(\alpha),\ x \simeq \phi(x),\ y \simeq \phi(y),\ S_\alpha \simeq S_{\phi(\alpha)},\ I_x \simeq I_{\phi(x)},\ 0_y \simeq 0_{\phi(y)},$$

where "≃" indicates set isomorphism.

We can proceed now to define the "logical spaces" suggested by von Neumann's work on self-reproducing automata. Our object is a single composition with two basic properties: (i) "universality", in the sense that any finite

-28-

OBJECT $\underline{A}_\gamma$

$X' = \{x_{12}, x_{21}, x_{31}\}$ $\qquad$ $Y' = \{y_{11}, y_{22}, y_{31}\}$

y	y_{11}	y_{22}	y_{31}
$\gamma(y)$	x_{21}	x_{31}	x_{12}

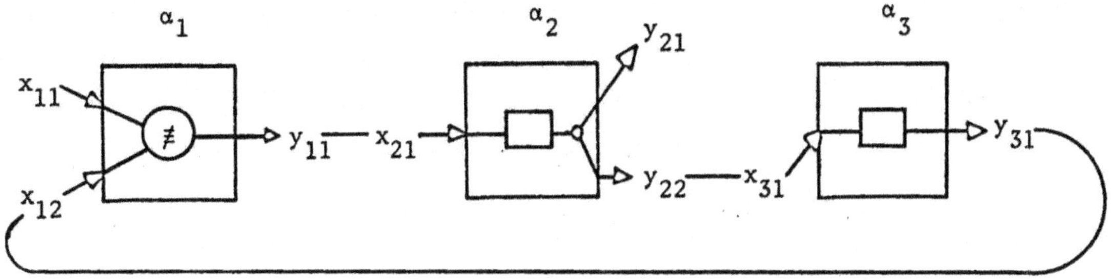

IMAGE $\underline{B}_{\gamma_1}$

$X'_1 = \{{}^1x_{11}, {}^1x_{21}, {}^1x_{32}, {}^1x_{41}, {}^1x_{61}\}$ $\qquad$ $Y'_1 = \{{}^1y_{11}, {}^1y_{21}, {}^1y_{32}, {}^1y_{41}, {}^1y_{51}, {}^1y_{61}\}$

${}^1y$	${}^1y_{11}$	${}^1y_{21}$	${}^1y_{32}$	${}^1y_{41}$	${}^1y_{51}$	${}^1y_{61}$
$\gamma_1({}^1y)$	${}^1x_{41}$	${}^1x_{11}$	${}^1x_{21}$	${}^1x_{51}$	${}^1x_{61}$	${}^1x_{32}$

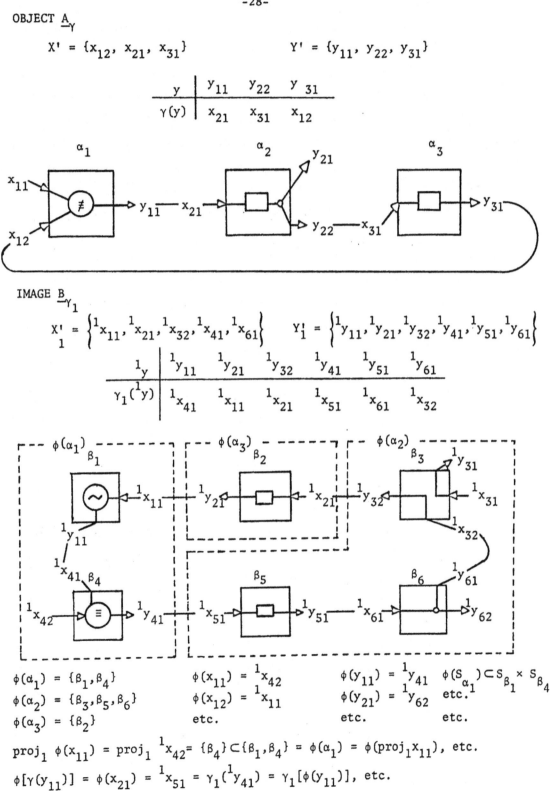

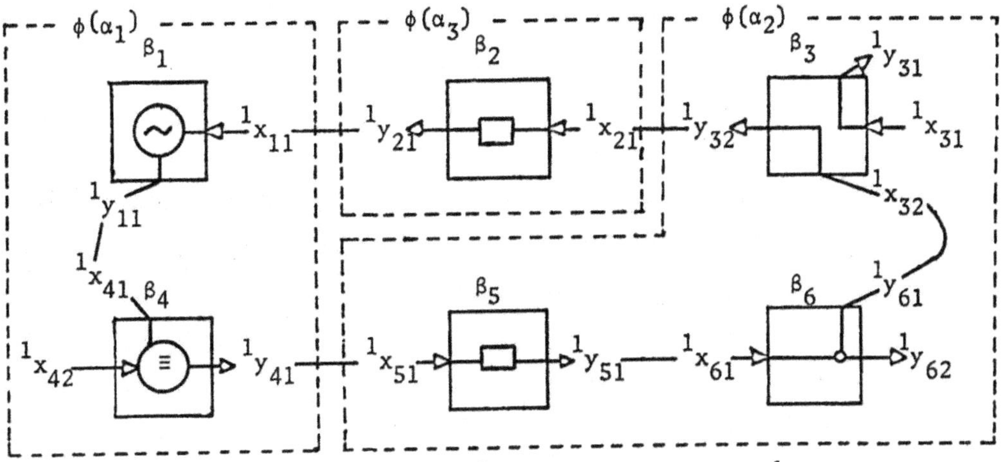

$\phi(\alpha_1) = \{\beta_1, \beta_4\}$ $\qquad$ $\phi(x_{11}) = {}^1x_{42}$ $\qquad$ $\phi(y_{11}) = {}^1y_{41}$ $\qquad$ $\phi(S_{\alpha_1}) \subset S_{\beta_1} \times S_{\beta_4}$

$\phi(\alpha_2) = \{\beta_3, \beta_5, \beta_6\}$ $\qquad$ $\phi(x_{12}) = {}^1x_{11}$ $\qquad$ $\phi(y_{21}) = {}^1y_{62}$ $\qquad$ etc.

$\phi(\alpha_3) = \{\beta_2\}$ $\qquad$ etc. $\qquad$ etc. $\qquad$ etc.

$\text{proj}_1\ \phi(x_{11}) = \text{proj}_1\ {}^1x_{42} = \{\beta_4\} \subset \{\beta_1, \beta_4\} = \phi(\alpha_1) = \phi(\text{proj}_1 x_{11})$, etc.

$\phi[\gamma(y_{11})] = \phi(x_{21}) = {}^1x_{51} = \gamma_1({}^1y_{41}) = \gamma_1[\phi(y_{11})]$, etc.

composition can be embedded therein, and (ii) "homogeneity", in the sense that, given any two "regions" in the composition, any embedding procedure which works in one region will work in the other. A more precise statement of the latter property can be based upon the notion of one embedding being a "translate" of another:

Given a composition $\underline{D}_\delta$ having a single generator, an embedding ϕ' of $\underline{A}_\gamma$ in $\underline{D}_\delta$ will be called a <u>translate</u> of an embedding θ of $\underline{A}_\gamma$ in $\underline{D}_\delta$ if

(1) There is an isomorphic embedding θ, of the subcomposition indexed by $\phi(A)$ on the subcomposition indexed by $\phi'(A)$, such that $\theta\phi = \phi'$,

(2) for all $\alpha, \alpha' \varepsilon A$, any $\xi \varepsilon \phi(\alpha)$, $\xi' \varepsilon \phi(\alpha')$, and any connected sequence ρ {from ξ to ξ'} {from ξ to $\theta(\xi)$}, there exists a connected sequence ρ' {from $\theta(\xi)$ to $\theta(\xi')$} {from ξ' to $\phi(\xi')$} such that $\text{proj}_2\, \rho = \text{proj}_2\, \rho'$.

A composition $\underline{V}_\nu$ will be called <u>universal and locally homogeneous</u> (for the embedding of finite compositions) or, briefly, <u>universal</u> if:

(1) Given any finite composition $\underline{A}_\gamma$ there is an embedding ϕ of $\underline{A}_\gamma$ into $\underline{V}_\nu$.

(2) If ϕ embeds $\underline{A}_\gamma$ in $\underline{V}_\nu$ then, given arbitrary $\alpha \varepsilon A$, $\xi \varepsilon \phi(\alpha)$, and $\xi' \varepsilon V$, there exists a translate ϕ' embedding $\underline{A}_\gamma$ in $\underline{V}_\nu$ so that $\theta(\xi) = \xi'$; i.e., the same embedding procedure can be used to place the image anywhere within $\underline{V}_\nu$.

A subset of the class of iterative circuit computers

Holland, Iterative Circuit Computers".

---

satisfies the above definition, and hence establishes the existence of of universal compositions. Certain necessary conditions for a composition to be universal also follow immediately from the definition:

Lemma 2.1. If $\underline{V}_\nu$ is universal then

(1) V must be countably infinite,

(2) $\underline{V}_\nu$ must be generated by a single element g for which the output function u depends properly on both I and S (i.e., the generator is a proper Mealy automaton),

(3) strings over output indices must be "commutative", i.e., if ρ is a connected sequence from α to α', $\sigma = \text{proj}_2\rho$ and σ' is any permutation of σ, then there exists a connected sequence ρ' such that $\text{proj}_2\rho' = \sigma'$.

Part (2) of the lemma follows from the observation that there exist compositions with arbitrarily long connected sequences. The definition of an embedding requires that the image of any such composition contain a connected sequence at least equally long. (Note again that more than behavioral equivalence is required of the image). However, if the output function u_g of the generator g of $\underline{V}_\nu$ depends only upon S, a "delay" is imposed between input and output. As a consequence, in the image, the number of time-steps required to effect the transition function of the connected sequence will depend upon its length, which is unacceptable. Part (3) of the lemma follows from part (2) of the definition of a universal composition. By using a two element image and an appropriate translate of it, one can show that all strings of length two are "commutative"; induction on length then establishes (3).

Corollary 2.1. A universal composition can be "co-ordinatized" so that its elements appear at the intersections of a discrete (integer) cartesian k-dimensional grid (where $k \leqq n_\alpha$, the number of outputs of the generator).

(The proof of the corollary follows easily from part (3) of the lemma.) In terms of this corollary the second requirement in the definition assures that any translation of an image over the grid is also an image of the same object. As a consequence properties of the image, such as its connection scheme, can be made independent of its location in the grid.

While von Neumann's cellular space suggested the class of universal compositions, it is not itself a member of the class. The space is generated by a Moore-type automaton (u: $S \to O$) contradicting condition (2) of the lemma. It might seem that condition (2) could be relaxed enough to admit the von Neumann space if the compositions to be embedded themselves used only Moore-type elements. But this is not so even if a weakened form of embedding, a b-slow embedding, is used. A b-slow embedding lets input signals to the image occur at the reduced rate of once every b time-steps, i.e., at times bt; the states of elements and outputs in the image are then required to occur at the reduced rate of once every b time-steps, after an arbitrary finite initial transient period (which may vary from one composition to another), i.e., at times bt+c.

Theorem. Given any composition $\underline{W}_\omega$ generated by a Moore-type element, any finite set of automata G sufficient to generate all finite automata, and any integer $b \geqq 0$, there exist finite compositions $\underline{A}_\gamma$ generated by G which cannot be b-slow embedded in $\underline{W}_\omega$.

Proof outline:

The proof turns on the limited "packing density" of universal spaces, which in von Neumann spaces causes a rapid increase in propagation time.

As a consequence, for object compositions sufficiently large, the transition rate in the image falls behind the input rate.

It is easily established that, if a composition is to be universal for b-slow embeddings, all of the necessary conditions established in Lemma 2.1 apply, with the possible exception of the requirement that the single generator be a Mealy-type automaton. To show that this last requirement also applies, a contradiction will be developed from the assumption that the single generator can be a Moore-type element having a delay between input and output (a 'lag-time') $\tau_g > 0$.

Definition. The <u>separation</u> of two elements in a composition is the length of the shortest connected sequence between them.

Definition. The <u>diameter</u> of a composition is the maximum separation of its elements.

For any Δ there exists an ℓ such that some ('most') compositions of diameter ℓ or greater can only be embedded in a subcomposition of $\underline{W}_\omega$ of diameter at least $\ell + \Delta$.

> Given any set G sufficient to generate all finite automata, there are compositions of diameter ℓ with at least $2^{\ell+1} - 1$ elements. Lemma 2.1 applied to $\underline{W}_\omega$ implies that it can be co-ordinatized by a cartesian grid of some given dimension n. But then less than $(\ell')^n$ distinct elements can belong to any subcomposition of diameter ℓ' in $\underline{W}_\omega$.
>
> Choose ℓ so that $2^{\ell+1} - 1 \geq (\ell+\Delta)^n$.

Among the compositions with the property just described, there exist some (again, 'most') having a cycle which is both functionally dependent upon every composition input and only embeddable in $\underline{W}_\omega$ with diameter greater than $\ell+\Delta$.

An example is a composition having a cycle of diameter ℓ which includes all the elements in the composition. (See Figure 5.)

[Figure 5.]

The diameter of the cycle's image is $\geq \ell+\Delta$ since there exists a pair of functionally dependent elements in the image of separation $\ell+\Delta$ (because of the diameter of the image subcomposition) and both belong to the cycle.

Because of the functional dependence, the composition can not be embedded as two separate (independent) subcompositions; all elements in the image must be functionally connected.

Let τ_G be the maximum delay ('lag-time') associated with any element of the generating set G and choose $\ell \ni (\ell+\Delta) > \frac{b\ell\tau_G}{\tau_g}$; this condition on ℓ can always be satisfied because ℓ is a logarithmic function of Δ.

Select a composition $\underline{A}_\gamma$ satisfying all the foregoing conditions and in $\underline{A}_\gamma$ select a pair of elements α_1 and α_2 such that the images thereof have a separation at least $\ell+\Delta$. From the set of object sub-cycles containing this pair, select the one which yields the shortest image sub-cycle. α_1 will have (a unique) input x and output y belonging to the connected sequence defining that sub-cycle.

The required functional dependence in $\underline{A}_\gamma$ assures that $I_\gamma(t)|x$ depends in a non-trivial way on $O_\gamma(t')|y$ for some earlier t'. Since the object sub-cycle can be no greater than ℓ in diameter, the total delay from y to x over the connected sequence, $\tau(y,x)$, can be no greater than $\ell\tau_G$. (Briefly, the 'lag-time' between y and x cannot exceed $\ell\tau_G$). Hence $I_\gamma(t)|x$ depends upon $O_\gamma(t')|y$ for some $t' \geq t-\ell\tau_G$

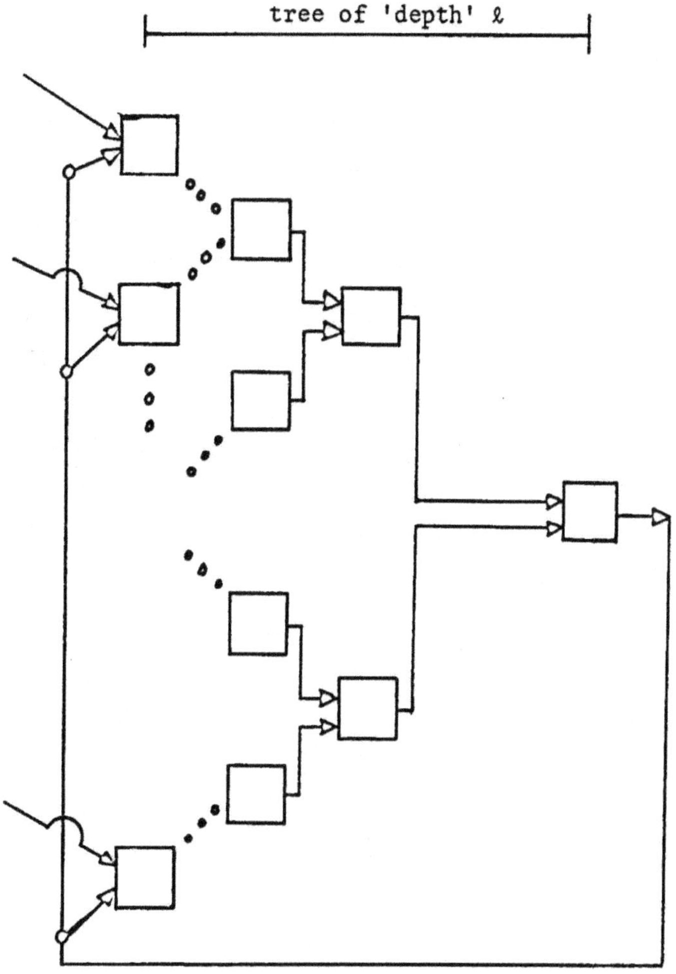

Figure 5. A Composition of $2^{\ell} - 1$ Elements with Cycle of Diameter ℓ.

(i) By the definition of a b-slow embedding, $\phi[I_\gamma](bt+c)|\phi[x]$ must then depend upon $\phi[O_\gamma](t')|\phi[y]$ for some $t' = b(t-\ell\tau_G) + c$.

(ii) However, the image subcycle has a diameter $\geqq \ell+\Delta$ yielding a total delay $\tau[\phi(y),\phi(x)] \geq (\ell+\Delta)\tau_g$; hence $\phi[I_\gamma](bt+c)|\phi[x]$ cannot depend upon $\phi[O_\gamma](t')|\phi[y]$ for any $t' = (bt+c) - (\ell+\Delta)\tau_g > bt+c - b\ell\tau_G = b(t-\ell\tau_G) + c$.

Statements (i) and (ii) contradict each other; hence compositions such as $\underline{A}_\gamma$ cannot be b-slow embedded in $\underline{W}_\omega$. It is not difficult to see that most compositions of sufficiently large diameter satisfy the conditions on $\underline{A}_\gamma$.

———— * ————

This theorem places strong restrictions on the use of von Neumann spaces for studies of self-applicable hierarchical descriptions and higher-level control in adaptive plans. The restrictions are emphasized by the following

<u>Corollary.</u> Given any finite composition $\underline{A}_\gamma$ and any integer $b \geqq 0$ such that $\underline{A}_\gamma$ can be b-slow embedded in $\underline{W}_\omega$ as above, 'most' compositions containing $\underline{A}_\gamma$ as a subcomposition cannot be b-slow embedded in $\underline{W}_\omega$.

Thus, given a set of computation procedures with a common "subroutine", it will in general be impossible to embed them in $\underline{W}$ so as to preserve the common subroutine.

The von Neumann space is, however, a member of a broader class, the computation-universal compositions, obtained by weakening condition (1) of the definition of universal compositions:

(1') Given any computable function Γ there is an embedding ϕ into $\underline{V}_\nu$ of a composition $\underline{A}_\gamma$ capable of computing Γ.

See Thatcher, "Notes on Turing Machines and Self-Description and on von Neumann Machines and Self-Reproduction" in Programming Concepts, Automata and Adaptive Systems, for an example of a definition of "computable" tailored to the present context.

In a universal space an embedded device is represented at any moment by a pattern of states assigned to a (usually contiguous) set of elements of $\underline{V}_\nu$. If the object device is a finite automaton, the set of elements will be indexed by a finite subset D of the index set V and will constitute a finite subcomposition $\underline{V}_{\nu|D}$ of $\underline{V}_\nu$. The pattern of states assigned to $\underline{V}_{\nu|D}$ will change to reflect changes of state in the object. But, if no single pattern in the image represents the structure of the object, how are we to modify this structure? The problem is complicated because several quite distinct devices can be embedded in any finite subcomposition of $\underline{V}_\nu$.

Formally, aspects of the structure can be extracted by constructing an appropriate equivalence class over the set of all states $S_{\nu|D}$ of the sub-composition $\underline{V}_{\nu|D}$. Define

$$\sim(s_1,s_2)_{[f_{\nu|D},I]}, \text{ for } s_1,s_2 \in S_{\nu|D} \text{ and } I \subseteq I_{\nu|D},$$

if there exist sequences σ_1 and σ_2 over I such that $f_{\nu|D}(\sigma_1,s_1) = f_{\nu|D}(\sigma_2,s_2)$. Then $\overset{\equiv}{}(s_1,s_2)_{[f_{\nu|D},I]}$ if and only if $\exists\, s_{j_1}, s_{j_2}, \ldots, s_{j_k} \in S_{\nu|D}$ such that $\sim(s_{j_h}, s_{j_{h+1}})_{[f_{\nu|D},I]}$, for h = 1, 2, ..., k-1, and $s_{j_1} = s_1$, $s_{j_k} = s_2$. If I is the image of the input alphabet of the object device, then a unique element of the equivalence class can be associated with each initialization (initial state assignment) of that device. That is, for any state accessible from the initial state, the associated element of the equivalence class will contain an image state satisfying all conditions for an embedding in $\underline{V}_{\nu|D}$.

Moreover, any initialized device which can be embedded in $\underline{V}_{\nu|D}$ with I as the image of its input alphabet can similarly be associated with some element of this equivalence class. A change in I will yield a different equivalence relation and a different equivalence class. Given any possible embedding of an initialized device in $\underline{V}_{\nu|D}$, it can be associated with an element of an appropriately chosen equivalence class. In these terms changes of structure can be associated with transformations between elements of (possibly different) equivalence classes.

How are these transformations to be managed within the space? From the requirements on embedding, we know that changes of state in the image are independent of signals on all input lines to $\underline{V}_{\nu|D}$ except those lines which are images of the object's lines. Thus, changes of structure must be effected by signals on the image lines, but signals which are not images of object signals. More precisely: Let $\phi(X-X')$ designate the indices of images of input lines to the object device. For an embedding it is only required that $\phi(I_X) \subsetneqq I\phi_{(X)}$ for each $x \varepsilon X-X'$. Thus it can easily be arranged that

$$I_c = \prod_{x \varepsilon X-X'} [I_{\phi(x)} - \phi(I_X)]$$

exists and produces desired transformations. Given any embedding in $V_{\nu|D}$, the set I_c contains the set of all signals capable of modifying the structure of the image. Such a modification may, of course, alter the set of input lines capable of affecting the state transitions in $\underline{V}_{\nu|D}$--this amounts to a transformation from an element of one equivalence class to an element of another (distinct) equivalence class.

3. Incorporation of Models

From the foregoing it is apparent that the signals from one embedded device may control the construction or modification of another embedded device. It is only necessary that the signals from the controlling device belong to the set I_c of the device to be modified. When this condition is satisfied, the controlling device can control the construction operations in much the same way a finite automaton controls operations on the tape of a Turing machine. Examples of embedded constructing automata can be found in Thatcher's paper referred to on page 36.

See also Codd, Propagation, Computation, and Construction in Two-Dimensional Cellular Spaces.

These devices, because of their essentially serial, step-by-step operations, are (relatively) simple in conception, though not in implementation. Parallel construction procedures, mimicing highly-parallel biological construction such as the chromosome-controlled development of a cell, are also easily conceived and (less easily) implemented.

See Burks, "Computation, Behavior and Structure in Fixed and Growing Automata"; Holland, "Outline for a Logical Theory of Adaptive System,"; and Myhill, "Self-Reproducing Automata" in Programming Concepts, Automata and Adaptive Systems.

There will be more to say about construction procedures shortly, with special reference to hierarchical descriptions, but before that it will pay to take a closer look at the advantages of universal spaces for studies of construction.

When used to study embedded devices, a universal space is an analogue of the "spaces" used in physics to study its various "mechanics". With a physical space we associate a geometry (e.g. Euclidean geometry) and a set of state-transition laws holding without change at each point in the geometry (e.g. Newton's laws). It is the task of the theoretical physicist to derive the properties of structures embedded in this space (e.g. motions in a central field). From Corollary 2.1 we know that the universal space can be co-ordinatized, thus giving it a (discrete) geometry. By Lemma 2.1 we know that the space must be generated by a single element; thus the transition and output functions of the generator apply at each point, determining behavior accordingly. The result is a kind of discrete physics enabling us to concentrate on the information-processing properties of the embedded devices (in contradistinction to the physicist's primary interest in energy transformations).

One immediate value of this space, mentioned earlier, is an assurance that structures embedded in the space are LECF. Just as the laws of a physical space insure consistency of physical transformations so the transition rules of a universal space insure that any sequence of construction operations executed within the space preserve consistency of the image composition functions. This follows immediately from the requirement that the universal space be a composition--its behavior, and hence that of anything embedded therein, must always be well-defined by the definition of composition. Thus, any sequence of construction operations which can be carried out in the space must transform a device (composition) into a device (composition).

It is a consequence of the geometry of a universal space that only a limited number of modifications and additions can be made per time-step on any given (embedded) device. The limit is set by the number of elements (copies of the space's generator) occupied by the device and potential additions

That is, the amount of construction which can be carried out is dependent upon the size of what is already there--this contrasts sharply with a definition which would permit a construction sequence to be an arbitrary effective (recursive) function over the set of finite automata.

See Wang, "Circuit Synthesis by Solving Sequential Boolean Equations."

Tying the construction process to available resources has important consequences for adaptation. For example, substitution instances for schemata must be drawn, primarily, from a "pool" of already constructed devices or from limited modifications thereof. This requirement amounts to a "simplicity" ordering on the instances tried out by the adaptive plan--an ordering underlying that described earlier. As mentioned there, simplicity orderings play a central role in inductive inference, which is after all the adaptive plan's main task.

The geometry, by assigning each schema to a volume in the (n-dimensional) universal space, has another less desirable consequence. Suitable empty regions must be allocated or prepared to receive the devices which complete the schema. This raises problems of "shape" and "fit" which may be of interest only in quite detailed studies. It is possible to tackle these problems directly; one can set up control devices which, when necessary, change the size of the empty "substitution" regions in the embedded schema, etc.

See Newell, "On Programming a Highly Parallel Machine to be an Intelligent Technician."

However for many purposes, both practical and theoretical, it may be simpler to code the hierarchical descriptions first as a string and then operate thereon. (This is reminiscent of the encoding of three-dimensional protein molecules as one-dimensional strings on the chromosomal helices). "Making room" in a string for the insertion of another string is a much simpler operation than the corresponding operation in two or more dimensions. (The "genetic" operators such as crossover, inversion, etc., provide a ready repertory of string manipulators suited to the present requirements). Any of the standard techniques for embedding a tree structure in a sequential array

See McCarthy, "Recursive Functions of Symbolic Expressions and their Computation by Machine, Part I."

will serve for reducing the skeleton of the hierarchy to a one-dimensional form. The labels of vertices, other than the terminal ones, are already strings. The terminal instances will either be drawn from the list of primitives or from the pool of descriptions of already constructed devices. Thus if the primitives can be given in string form the whole hierarchical description can be given as a string. This requirement can be met in any of several ways; for example, the primitives can be specified by subroutines for a universal Turing machine or some other general-purpose device.

See Holland, "Outline for a Logical Theory of Adaptive Systems".

(There are of course procedures for making this coding better attuned to genetic operators, but this is a lengthy subject not central to the present discussion).

If the hierarchical description is encoded in string form, construction becomes a two-step procedure. First the string must be modified to reflect the result of the construction operation. Then the resulting string will have to be read off by an (embedded) "computer" which can either simulate the device described (in the fashion of a universal Turing machine) or else translate the string into an embedded device (in the fashion of a von Neumann constructor). Only in the latter case will there be an immediate test of the modified description to see if it retains the LECF property--if the string can be translated the result is perforce LECF since it lies within the space.

There are two classes of construction operations on hierarchical descriptions: those which rearrange the hierarchical description to obtain a new description of the same device, and those which yield a description of a new device. Among the most important operations of the first type (for the prototype hierarchies given earlier) are those for changing cable configurations (splitting or coalescing cables and reordering wires within cabels) and those for rearranging the hierarchical skeleton (by coalescing levels or distinguishing new levels). Such changes have a strong influence on the schemata likely to be generated and tested by the adaptive plan. Among the most important operations of the second type are those which delete vertices to yield schemata and those which substitute other descriptions at terminal vertices (including devices which act as alphabet or time-scale "translators", modifying interface conditions). The details of these construction operations will of course depend heavily upon the class of hierarchical descriptions and upon the encoding. There are several convenient overall organizations, all more or less routinely implemented. (One of the most interesting, following a genetic format, would employ hierarchies somewhat different than the earlier prototypes).

Rather than looking at a detailed (and rather routine) example at this point, let us push on to some of the broader questions of implementation and usage.

When discussing substitution in schemata we talked of drawing upon a pool of already constructed devices. Treated properly, this pool can serve as the plan's repository of information about the outcomes and evaluations of previous trials. For instance, the plan can use the pool in such a way that it generates a probability distribution over schemata, a distribution permitting inferences about the performance of schemata (as discussed in Section 2). One way the plan can do this is to treat the descriptions of devices in the pool as a population of individuals undergoing a process of recombination. The plan as applied to the population can then take the following general form:

(1) In preparation for the recombination process, each description is copied a number of times determined by the corresponding device's past performance. That is, each description can be thought of as producing "offspring" in accordance with its performance in the environment confronting the adaptive plan. (As an example, the number of copies may be a random variable having a mean determined by the performance measure).

(2) In the resulting population, descriptions are grouped in sets (singletons, pairs, etc.) appropriate for the application of particular recombination operators. The operators are then applied to these sets to produce the corresponding exchanges and rearrangements of parts.

Of course, this plan could equally well be taken as a general description of the processes of population genetics. Looking to that analogy, we see that it can easily be arranged that the pool of devices emerging from step (2) has a negligible intersection with the pool of devices presented to step (1). (Barring identical multiple births, no two humans have the same genetic description.) Thus the plan satisfies the desideratum that it be able to proceed where necessary with negligible duplication of trials. At the same time, various schemata will appear multiply, as parts of several descriptions, and will be sampled accordingly. The proportion of a given schema in the overall population will depend upon two factors: the average performance of its instances and its <u>dispersion</u> or generality. (Roughly, the dispersion of a schema increases with the brevity of its hierarchical description. If the string descriptions are formed over a finite alphabet of k letters and, for example, if all combinations of letters are equally likely, then the probability of finding a given schema with m letters in its description will be k^{-m}. While this factor will change as the probability distribution is skewed from uniform, it is clear that the lower the dispersion of a schema the rarer it will be in general.) The net effect of this plan is a time-dependent probability distribution over schemata, conditioned on past performance and generality. Thus, the population summarizes the history of the adaptive plan's confrontation with the environment. In other words, the pool of devices available to the plan at any given time constitutes the current state of its knowledge of the environment, as intended.

If we are to use a universal space to study this plan, the plan must somehow be put into effect within the confines of the space. How is this to be accomplished? It is clear (because a universal space is countably infinite and homogeneous) that any number of devices, acting simultaneously, can be embedded in the space at any time. Thus the pool of devices (and their descriptions) can be placed in the space.

It is perhaps less clear that arbitrary devices can be embedded so that they can be shifted from point to point without disruption. Yet this seems the most natural way to handle the groupings required by step (2) of the plan (particularly if we want to execute each step of the plan in a parallel rather than a serial fashion). The possibility of a "shift" operation rests on the second requirement in the definition of a universal composition: given two points in the space, and an embedded device arranged around one of the points, there exists a translate of that device similarly arranged around the other point. If the two points are adjacent, then corresponding points in the two images will be directly connected (by outputs of the same index). Now, let the generator of the space be so chosen that the state of the generator can be transmitted over (some) of its outputs (for some $y \in Y'$, $O_y \supseteq S$). If each of the elements constituting the support

---

See Thatcher, "Notes on Turing Machines and Self-Description and on von Neumann Machines and Self-Reproduction" in Programming Concepts, Automata and Adaptive Systems.

---

of the embedded device simultaneously receives a signal causing its state to be transferred over output y, a kind of "shift" operation ensues. The result is a translation of the image within the space. (With some care, it can be arranged that the "shift" signal originates internally when certain conditions are encountered on free inputs of the image subcomposition-- the embedded device is then "self-propelled"). Attaining simultaneity of the "shift" signal throughout the image is easily solved in certain iterative circuit computer spaces

---

See Holland, "Outline for a Logical Theory of Adaptive Systems".

---

although it leads to a pretty problem in von Neumann's space. (See Moore's discussion

---

See Moore, "The Firing Squad Synchronization Problem."

---

of the "firing squad" problem; the problem arises in any computation universal space which is not universal). Once there is provision for shifting images, it is no great problem to provide the groupings required by step (2) of the proposed adaptive plan.

On this basis we can proceed to implement the remainder of the plan. If the implementation is carried out in an iterative circuit computer, the procedures required can be programmed much as one would write general subroutines for a digital computer. One such version can be set up along the following lines:

(i) The embedded set of descriptions is used to construct the corresponding pool of devices (as outlined on page 41).

(ii) The devices are tested against the environment and the description corresponding to each device is copied a number of times determined by its performance (as outlined on page 43--if performance is determined by payoff and if the number of copies is to be a random variable, then the payoff level is used to set the mean of an associated pseudo-random number generator).

(iii) The descriptions undergo simultaneous random walks within some region of the space set off by reflecting barriers. (This involves the shift operation discussed on page 45, the direction being controlled by a pseudo-random number generator associated with the description).

(iv) After a period long enough to assure thorough mixing, the descriptions are allowed to pair on contact (but no more) for a period long enough to yield some expected number of pairs. (That is, steps (iii) and (iv) together yield a set of randomly paired descriptions together with a "remainder" set of randomly selected singletons--more could be done, giving the grouping process a much more controlled aspect, but this suffices).

(v) The descriptions, paired and unpaired, undergo recombination; each description has an associated subroutine which (conjointly with its partner, when it is paired) selects (perhaps stochastically) and executes an appropriate recombination operator. (For example, a cross-over-like process could only be executed on pairs--in its simplest form it would amount to aligning the two strings, selecting randomly a length less than the minimal length of the two strings, and exchanging the initial segments of this length. Note that the recombination subroutine could itself be attached as a description to the device description, being used as in step (i) to yield the actual recombination subroutine. The recombination portion of the description could then also undergo recombination, permitting selection of the associated recombination operations on the basis of performance. This procedure could in turn be compounded, automatically, to yield control hierarchies of whatever form proves advantageous vis-a-vis the particular environment confronting the adaptive plan).

(vi) Return to step (i).

There are of course many specific ways of filling in this outline, each replete with an overwhelming amount of detail. In those cases to date in which the details have been filled in, they seem to contribute little to a deeper understanding of the relations between hierarchical structures, cellular spaces, and adaptation. (What understanding is gained is primarily that gained in the working out, rather than in the result; if the working out is a model programmed for simulation, the sumulation itself can be quite suggestive of new paths.)

See Codd, Propagation, Computation, and Construction in Two-Dimensional Cellular Spaces; Rosenberg, Simulation of Genetic Populations with Biochemical Properties; Bagley, The Behavior of Adaptive Systems which Employ Genetic and Correlation Algorithms for relevant results.

I will accordingly once again set aside the burden of detail in order to pursue some more general issues.

Under the foregoing arrangement, information initially supplied to the adaptive plan yields the set of schemata prominent in the initial corpus. These are treated by the plan as primitives out of which to construct better-adapted structures (schemata). Schemata which are successful will persist (occur with substantially higher than average probability) because of a higher than average duplication rate. Thus they have a greater chance of serving as the components of more complex schemata. Generally, the constructed schemata will appear in a significant number of still more complex schemata only if they are more successful than their components. A natural hierarchy emerges. The blocks at the lowest level of the tree are schemata, or instances of schemata, formed by a relatively few operations on the initial corpus and persistent under successive iterations of the adaptive plan. The higher-level blocks are those, formed in turn by relatively few operations on lower-level blocks, which are also persistent under successive iterations. Without the suggested hierarchical organization, it would take much longer for devices of a similar number of primitives and of comparable success to emerge. For example, assume as on page 44 that some description is at least m letters long on an alphabet of k letters. Then, in the absence of the suggested hierarchical structure, the plan can expect to construct k^m trials before it first encounters the device. Stated in an intuitive but slightly misleading way: There is only time enough for devices with hierarchical organization to emerge under the adaptive plan's guidance. It is useful to compare this hierarchy to the analogous natural hierarchies of stability in open chemical systems, and to the organelle-cell-tissue-organ-organism-species-...hierarchies, paying particular attention to the increase in half-life as one moves up the hierarchy.

Perhaps the most significant feature of this plan is the fact that established schemata, in effect, become new primitives based on the plan's accumulated information about the environment. These new primitives give the plan new ways of representing aspects of the environment. Such changes in representation can be vital in taking advantage of regularities in the environment; indeed such regularities often are revealed only when appropriately represented.

4. Concluding Remarks

The subject of the present discussion is only a preliminary to the investigation of adaptive systems, and it should be recognized as such. Parts of it may have intrinsic formal interest, but in my eyes it will have failed of its objective if it offers no help in resolving questions about adaptation. For such purposes a formal framework and its attendant apparatus are to be tolerated only if they enable answers to be obtained for questions of prior interest--questions originating outside the formalism. In the case of adaptation these questions all center upon the notion of efficiency. At first sight, the concept of efficiency seems far removed from the formal definition of structure, but some of the connections are indicated in the last few paragraphs above. ("Persistence" and "enough time ... to emerge" are concomitants of efficiency.) Answers to similar questions were also von Neumann's ultimate objective in his unfinished work: "...can the construction of automata by automata progress from simpler types to increasingly complicated types? Also, assuming some suitable definition of 'efficiency', can this evolution go from less efficient to more efficient automata?"

---

von Neumann, Theory of Self-Reproducing Automata.

Here, as elsewhere, the overall objective is a formalism sufficiently oriented to reality to permit reliable inference, at least qualitatively, about what would happen in the more complex real situations. (The situation is quite like that of the use of "free fall", in Newtonian physics, as a guide to more complex cases involving, say, atmospheric friction). When so conceived, a formalism has striking advantages as a tool augmenting empirical investigation. Unlike the real situation, actions within the formal system are completely defined and available for deductive analysis or simulation. The universal spaces are examples par excellence, yielding completely defined universes within which one can embed models of adaptive processes. This largely eliminates the great pitfall of informal theories: placing too great a burden on some vague or ill-defined mechanism which cannot bear the brunt (e.g. a mechanism which cannot possibly act consistently as required). Once an adaptive plan is presented within the formalism its consistency is assured, along with its formal existence, and one can test inferences about it under conditions admitting of no hidden confounding factors.

REFERENCES

1. Burks, A. W., "Computation, Behavior and Structure in Fixed and Growing Automata" in Self-Organizing Systems, Pergamon Press, 1960, p. 282-311.

2. Burks, A. W. and Wright, J. B., "Theory of Logical Nets", Proc. IRE 41, 1953, 1357-1365.

3. Bagley, J. E., The Behavior of Adaptive Systems which Employ Genetic and Correlation Algorithms, The University of Michigan Ph.D. Dissertation, 1967.

4. Codd, E. F., Propagation, Computation, and Construction in Two-Dimensional Cellular Spaces, The University of Michigan Ph.D. Dissertation, 1967.

5. Holland, J. H., "Iterative Circuit Computers", Proc. Western Joint Computer Conference 1960, 259-265.

6. Holland, J. H., "Outline for a Logical Theory of Adaptive Systems", J. Assoc. Computing Machinery 9, 1962, 297-314.

7. Holland, J. H., "Universal Spaces: A Basis for Studies of Adaptation" in Automata Theory, Academic Press, 1966, p. 218-230.

8. McCarthy, J., "Recursive Functions of Symbolic Expressions and their Computation by Machine, Part I", Comm. Assoc. Computing Machinery 3, 1960, 184-195.

9. Moore, E. F., "The Firing Squad Synchronization Problem" in Sequential Machines: Selected Papers, Addison-Wesley, 1964, p. 236-237.

10. Myhill, J., "Self-Reproducing Automata" in Programming Concepts, Automata, and Adaptive Systems, The University of Michigan Summer Conferences, 1966.

11. Newell, A., "On Programming a Highly Parallel Machine to be an Intelligent Technician", Proc. Western Joint Computer Conference, 1960, 267-282.

12. Putnam, H., Probability and Confirmation, Forum Lectures, Voice of America, U.S.I.A.

13. Rosenberg, R. S., Simulation of Genetic Populations with Biochemical Properties, The University of Michigan Ph.D. Dissertation, 1967.

14. Samuel, A. L., "Some Studies of Machine Learning, Using the Game of Checkers I and II-Recent progress", IBM J. Res. and Dev. 3, 211-229 and 11, 601-617, 1959 and 1967.

15. Thatcher, J. W., "Notes on Turing Machines and Self-Description and on von Neumann Machines and Self-Reproduction" in Programming Concepts, Automata and Adaptive Systems, The University of Michigan Summer Conferences 1966.

16. Von Neumann, J., _Theory of Self-Reproducing Automata_ (edited and completed by A. W. Burks), University of Illinois Press, 1966.

17. Wang, H., "Circuit Synthesis by Solving Sequential Boolean Equations", _Zeitschift f. Math. Logic u. Grundlagen d. Math. 5_, 1959, 391-322.

THE UNIVERSITY OF MICHIGAN
College of Literature, Science, and the Arts
Computer and Communication Sciences Department

Technical Report

ADAPTIVE PLANS OPTIMAL FOR PAYOFF-ONLY ENVIRONMENTS

John H. Holland

Supported By

Department of Health, Education, and Welfare
National Institutes of Health
Grant No. GM-12236-03
Bethesda, Maryland

Department of the Navy
Office of Naval Research
Contract No. N00014-67-A-0181-0011
Washinton, D. C.

and

U. S. Army Research Office (Durham)
Grant No. DA-31-124-ARO-D-483
Durham, North Carolina

Administered Through

OFFICE OF RESEARCH ADMINISTRATION ANN ARBOR

May 1969

ABSTRACT

This paper characterizes a class of adaptive algorithms, the _reproductive plans_, which produce optimal performance in conditions where the information fed back to the algorithm consists only of a payoff at each instant of time. The payoff function can be any bounded (non-linear) function of the algorithm's output space. The reproductive plans have the advantage that they achieve a global optimum (over the whole time course) via a local step-by-step optimization of well-defined quantities. The theorem guaranteeing optimal performance is a modification of the Kuhn-Tucker fixed point theorem closely related to Gale's work in mathematical economics.

Adaptive Plans Optimal for Payoff-Only Environments

by

John H. Holland

Dept. Computer and Communication Sciences

The University of Michigan

A wide variety of adaptive processes can be usefully described in a framework consisting of three basic elements: the adaptive plan proper, a set of admissible environments, and a criterion of improvement. The plan τ works with a repertory $\mathscr{A}$ of procedures, devices or techniques which it can bring to bear on the environment E in its attempt to adapt. In the cases of most interest τ uses a set of operators Ω to generate new elements of $\mathscr{A}$ for testing against the environment E. Information accumulated from previous tests of E determines which elements are generated and tested at each step; as a consequence τ responds differently to different environments. The set of admissible environments $\mathscr{E}$ is the domain of environments in which the adaptive plan may be required to act; $\mathscr{E}$ may be equally well looked upon as a characterization of what is initially unknown to τ. The criterion of improvement χ provides a standard whereby different adaptive plans from some set of alternatives $\mathscr{T}$ can be compared and ranked. Typically the criterion will relate to some intuitive notion of optimality such as efficiency, stability, fitness, etc.

As an illustration of this $\langle \mathscr{E}, \mathscr{T}, \chi \rangle$ framework consider game-learning programs: $\mathscr{A}$ consists of various strategies for play of the game, such as Samuel's polynomial functions $\Sigma a_i \vartheta_i$; Ω consists of various operators for interchanging or modifying the weights a_i; each $\tau \in \mathscr{T}$ is a prescription for employing the operators on the basis of the outcome of plays of the game ; $\mathscr{E}$ characterizes the set of payoff assignments the plan is prepared to deal with; and χ measures how rapidly a given plan improves its performance. A multitude of other problems in optimal control, genetics, game-playing, mathematical economics, pattern recognition, etc., readily fit this framework.

In this paper attention will be restricted to $<\mathcal{E}, \mathcal{T}, \chi>$ processes wherein an environment $E \in \mathcal{E}$ is characterized by a function $\mu_E: \mathcal{A} \to R$, to be interpreted as an assignment of payoff to each $A \in \mathcal{A}$. It will further be assumed that, when a plan $\tau \in \mathcal{T}$ tests a procedure A against E, it only receives or employs information about the payoff $\mu_E(A)$. $<\mathcal{E}, \mathcal{T}, \chi>$ processes so-restricted will be called first-order processes. The performance of an optimal first-order plan τ^* sets a non-trivial lower bound on the performance of any plan which receives additional information, such as information about the state of the environment or the actions of the trial devices therein. Finally, although it is primarily a convenience, it will be assumed that $\mathcal{A}$ is finite, of cardinality m, and ordered. Thus each μ_E can be treated as a vector of m components, component i being the real assigned by μ_E to the ith element of $\mathcal{A}$.

The initial state of the adaptive plan will be given by assigning a probability to each element of $\mathcal{A}$, thus producing a sample space based on $\mathcal{A}$. Using the ordering of $\mathcal{A}$ again, this assignment can be looked upon as a vector p_0 from the set $\mathcal{P}_m = \{p = (p_1, \ldots, p_i, \ldots, p_m) \quad 0 \leq p_i \leq 1 \text{ and } \Sigma_{i=1}^m p_i = 1\}$. The adaptive plan τ will operate by modifying the sample space on the basis of payoff information received from an initial sampling. Since $\mathcal{A}$ is fixed, this means that the probabilities assigned to $\mathcal{A}$ will be modified. The plan's options will be determined by a set of operators Ω for transforming the probability assignments. Each operator $\omega \in \Omega$ will be defined by a set $\{(p, p')\}$ of pairs of probability vectors drawn from $\mathcal{P}_m$, the first element p of each pair designating a possible input to the operator, and the second element p' designating the result of the operator's action on that input. For uniform format, the operators and the corresponding sets of pairs will be ordered, so that there is a natural order on the collection of all vector pairs $\{(p_1, p_1'), \ldots, (p_n, p_n')\}$ corresponding to μ. Thus μ can be represented by a pair of matrices (W,W') where the jth column of W is just the vector p_j and the jth column of W' is p_j', $j = 1, \ldots, n$.

The class of plans $\mathcal{T}$ can be defined in a natural way by permitting plans to employ mixtures of operators, applying each operator to each point in the sample space with a given probability. Thus, the action of a plan $\tau \in \mathcal{T}$ at any point in time can be specified by a probability vector over the collection of pairs. Formally, $\mathcal{T} = \{$ sequences $\tau = \langle\tau_t\rangle$, $t = 0,1,2,\ldots$ (i) $\tau_t \in \mathcal{P}_n$, (ii) $W\tau_0 = P_o$, (iii) $W'\tau_t = W\tau_{t+1}\}$. Condition (ii) requires that the initial sample space specified by p_0 serve as an input to the mixture of operators specified by τ_0; condition (iii) requires that the output of each time-step serve as the input for the next.

To define a criterion χ, first note that $W'\tau_t$ defines the sample space over $\mathcal{A}$ resulting from the adaptive plan's operations at time t. Accordingly $\mu_E W'\tau_t$ gives the expected value of a sample drawn from this space at time t. Thus with plan $\langle\tau_t\rangle$ we can associate the sequence of expected payoffs $\langle\mu_E W'\tau_t\rangle$. A plan $\tau^* = \langle\tau_t^*\rangle$ is defined to be E-optimal if

$$(\forall \tau \in \mathcal{T}) \; [\liminf_{T \to \infty} (\Sigma_{t=1}^{T} \mu_E W'\tau_t^* / \Sigma_{t=1}^{T} \mu_E W'\tau_t) \geq 1].$$

We can now go to the key result: The optimal expected payoff sequence for any environment E of a first-order process can be produced by a plan of a very particular type, a reproductive plan, which at each time-step maximizes a function of quantities defined for that time-step. That is, the global optimum is obtained by iteration of a local optimization procedure. Stated yet another way, for any problem in adaption which can be presented as a first-order process, there is a reproductive plan which solves the problem by iterated maximization of a locally defined quantity.

The proof can be approached as follows: Informally, a reproductive plan can be looked upon as acting on a population of elements from $\mathcal{A}$, say a typical sample from the sample space. At each time-step the plan modifies the population by operating upon it in two phases. During the first phase, the reproductive phase, each member of the current population is duplicated a number of times dependent upon the payoff it receives. During the second phase, the recombination phase, the plan subjects the members of the duplicated population to decomposition and recombination using the operators in Ω. Formally, the reproductive plans can be defined as elements of the set $\mathcal{T}'$ = { vector sequence pairs $(\langle m_t\rangle, \langle \tau_t'\rangle)$, $t = 0,1,2,\ldots \ni$ (i) $W\tau_0' = d_{m_0} p_0$, (ii) $W\tau_{t+1} = d_{m_t} W' \tau_t$ where d_{m_t} is the diagonal matrix with the components of m_t on the diagonal }.

Lemma. Assuming the rank of W equals the dimension of $\mathcal{P}_m$, there exists a sequence of diagonal matrices $\langle d_t\rangle$ such that for any $\langle \tau_t'\rangle \varepsilon \mathcal{T}'$, $\tau_t' = (\Pi_{t'=1}^{t} d_{t'})\tau_t$ where $\langle\tau_t\rangle \varepsilon \mathcal{T}$, and vice-versa.

Theorem. For E ε $\mathcal{E}$ of a first-order process such that the rank of W equals the dimension of $\mathcal{P}_m$: $\langle\tau_t\rangle \varepsilon \mathcal{T}$ is optimal $\Leftrightarrow$ there exists $(\langle m_t\rangle, \langle \tau_t'\rangle) \varepsilon \mathcal{T}'$ such that τ_t' maximizes $\mu_{Et}^{o} + \Delta_{Et}^{o}$ at each time step, where by definition

$$\mu_{Et}^{o} = \mu_E W'(\Pi_{t'=1}^{t} d_{t'}^{-1})\tau_t', \text{ the discounted utility of } \tau_t',$$

$$\Delta_{Et}^{o} = m_{t+1} W(\Pi_{t'=1}^{t+1} d_{t'}^{-1})\tau_{t+1}' - m_t W(\Pi_{t'=1}^{t} d_{t'}^{-1})\tau_t', \text{ the discounted offspring advantage of } \tau_t'.$$

Proof outline:

(1) The sequence $\langle\mu_{Et}^{o}\rangle$ produced by $(\langle m_t\rangle, \langle\tau_t'\rangle)$ is identical to the sequence $\langle\mu_E W'\tau_t\rangle$ produced by $\langle\tau_t\rangle = \langle(\Pi_{t'=1}^{t} d_{t'}^{-1})\tau_t'\rangle$.

(2) Because $\mu_E W'\tau_t$ is a concave function of τ_t, Gale's version ("A Mathematical Theory of Optimal Economic Development", Bull. Amer. Math. Soc. 74 (2) 1968) of the Kuhn-Tucker fixed point theorem applies, yielding:

$$\langle\tau_t\rangle \text{ optimal} \Leftrightarrow (\exists\langle m_t\rangle)\ (\forall p \varepsilon \mathcal{P}_m)\ [\mu_E W'\tau_t + (m_{t+1}W' - m_t W)\tau_t \geq \mu_E W'p + (m_{t+1}W' - m_t W)p].$$

(3) $\mu_{Et}^{o} + \Delta_{Et}^{o} = \mu_E W'\tau_t + m_{t+1}W\tau_{t+1} - m_t W\tau_t = \mu_E W'\tau_t + (m_{t+1}W' - m_t W)\tau_t$.

(4) Thus, $\tau_t' = (\Pi_{t'=1}^{t} d_t)\tau_t$ maximizes $\mu_{Et}^{o} + \Delta_{Et}^{o} \Leftrightarrow \tau_t$ maximizes $\mu_E W'p + (m_{t+1}W' - m_t W)p \Leftrightarrow \tau_t$ belongs to an optimal plan $\tau \varepsilon \mathcal{T}$, as required.

A NEW KIND OF TURNPIKE THEOREM[1]

BY JOHN H. HOLLAND

Communicated by W. Givens, July 10, 1969

Briefly and informally an (economic or adaptive) *plan* can be described in terms of its action with respect to two sets:

$\mathcal{A}$, an effectively defined set of objects;

M, a set of functions $\mu_E: \mathcal{P} \to \mathcal{R}$, $E \in \mathcal{E}$, where $\mathcal{P}$ is the set of distributions over $\mathcal{A}$, and $\mathcal{R}$ is a ranking set of positive real numbers.

Under the intended interpretation, each $P \in \mathcal{P}$ corresponds to a mix of goods, chromosomes, strategies, or programs, $\mathcal{E}$ is a set of possible conditions or environments under which the plan is expected to operate, and the ranking of P, $\mu_E(P)$, specifies utility, expected offspring, payoff, or efficiency of the mix in condition or environment $E \in \mathcal{E}$. The plan, then, is a procedure (cf. sequential sampling procedure, dynamic programming policy) for searching $\mathcal{P}$ in an attempt to locate mixes of high rank in any given $E \in \mathcal{E}$; the object is to construct (if possible) a plan which is "robust with respect to M" in the sense that the search proceeds "efficiently" for any $\mu_E \in M$.

More formally, with any pair (τ, E) where τ is a plan and $E \in \mathcal{E}$, one can associate a trajectory through $\mathcal{P}$, $\langle \mathcal{P}(\tau, E)\rangle = \langle \mathcal{P}_1(\tau, E), \mathcal{P}_2(\tau, E), \cdots, \mathcal{P}_t(\tau, E), \cdots \rangle$; coordinated with the trajectory is the sequence of rankings $\langle \mu_E(\tau)\rangle = \langle \mu_{E,1}(\tau), \mu_{E,2}(\tau), \cdots, \mu_{E,t}(\tau), \cdots \rangle$ where $\mu_{E,t}(\tau) = \mu_E(\mathcal{P}_t(\tau, E))$. A plan τ_0 will be called *good* in E relative to a set of plans $\mathfrak{I}$ if

$$\sum_{t=1}^{\infty} [\overset{*}{\mu}_{E,t} - \mu_{E,t}(\tau_0)] = N_E(\tau_0) < \infty$$

where

$$\overset{*}{\mu}_{E,t} = \operatorname*{lub}_{\tau \in \mathfrak{I}} \{\mu_{E,t}(\tau)\}.$$

(The sense of good employed here is essentially that used, for example, by mathematical economists.) To assure that all the elements indexed by $\mathcal{E}$ represent nontrivial problems vis-à-vis the set of plans $\mathfrak{I}$ requires that, for all $t' > t$, $\overset{*}{\mu}_{E,t'} \geq \overset{*}{\mu}_{E,t} > 0$. (I.e., after a given time t,

---

[1] This report presents one phase of research on adaptation carried out at the Logic of Computers Group at The University of Michigan; the group is currently supported by ONR (N0014-67-A-0181-0011), ARO (DA-31-124-ARO-D-483) and NIH (GM-12236).

Reprinted from the BULLETIN OF THE AMERICAN MATHEMATICAL SOCIETY, November, 1969, Vol. 75, No. 6, Pp. 1311-1317

there will always be a plan which can give a performance of better than zero rank for a period of time.) A plan τ_0 will be called $(\mathcal{E}, \mathfrak{I})$-good if the plan is good in every $E\in\mathcal{E}$ relative to $\mathfrak{I}$. Under interpretation, an $(\mathcal{E}, \mathfrak{I})$-good plan is one which performs well in any $E\in\mathcal{E}$ relative to the set $\mathfrak{I}$ of (admissible or possible) plans under consideration. In these more rigorous terms, then, the objective is to construct (if possible) a plan which is $(\mathcal{E}, \mathfrak{I})$-good.

The purpose of this note is to demonstrate that, under rather weak conditions on $\mathfrak{I}$, it is possible to constructively define an algorithm ρ which is $(\mathcal{E}, \mathfrak{I})$-good for any collection M of bounded continuous functions μ_E. The first of the conditions on $\mathfrak{I}$ requires that, for each $E\in\mathcal{E}$, there exists a subset of good plans, $\mathfrak{I}_E$, occurring with non-negligible density (initial probability) in $\mathfrak{I}$. Loosely, this condition will be violated, when $\mathfrak{I}$ is finite, only if $\overset{*}{\mu}_{E,t}$ is determined by different plans at different times; i.e., it will be violated only if the sequences $\langle\mu_E(\tau)\rangle$ "determining" $\overset{*}{\mu}_{E,t}$ are oscillatory. When $\mathfrak{I}$ is not finite, the condition can also be violated if the sequences determining $\overset{*}{\mu}_{E,t}$ occur with negligible density. Since goodness is defined relative to the set $\mathfrak{I}$, the requirement is easily met if, for given E, some plan or set of plans with positive density eventually "dominates" all others in $\mathfrak{I}$; i.e., they "determine" $\overset{*}{\mu}_{E,t}$ for all t large enough. The second of the conditions on $\mathfrak{I}$ requires that, for any given $E\in\mathcal{E}$ and any time t, the trajectories associated with any set of probability intervals for the $A\in\mathcal{A}$ be a measurable subset of $\mathfrak{I}$ with respect to the initial distribution P_0. That is, for any $E\in\mathcal{E}$ and t, expectations for the distributions associated with various mixtures of trajectories are defined.

The theorem establishing the goodness of ρ can be looked upon as a kind of "turnpike" theorem (an analogue of von Neumann's prototype) which holds over a very broad range of conditions. It can also be given a Bayesian interpretation (first noticed by my colleague B. P. Zeigler) treating the $\mathcal{P}_t(\tau, E)$ as hypotheses and the $\mu_{E,t}(\tau)$ as evidence; the theorem then becomes a statement about the rate of convergence of Bayes's rule.

(A different version of the theorem can be obtained by changing the premise as follows: For each $\tau\in\mathfrak{I}$ and $E\in\mathcal{E}$ let

$$\mu_E(\tau) = \lim_{T\to\infty} \sum_{t=1}^{T} \mu_{E,t}(\tau)/T$$

and define

$$\overset{*}{\mu}_{E,t} = \operatorname*{lub}_{\mathfrak{I}_+\in \mathrm{pos}_\epsilon(\mathfrak{I})} \ \operatorname*{lub}_{\tau\in\mathfrak{I}_+} \mu_E(\tau)$$

where $\text{pos}_\epsilon(\mathfrak{J})$ is the set of subsets of measure exceeding ϵ in $\mathfrak{J}$. Then there is a ρ such that

$$\sum_{t=1}^{\infty} [\mu^{*}_{E,\epsilon} - \mu_{E,t}(\rho)] = N'_{E,\epsilon}(\rho) < \infty.)$$

The algorithm ρ, which will be called a *reproductive plan*, can be defined as follows: Let the element $\mathcal{P}$ selected at t by ρ acting in $E \in \mathcal{E}$ be given in terms of the (observed) sequences $\langle \mu_{E,t'}(\tau) \rangle_{t'=1,\cdots,t}$, $\tau \in \mathfrak{J}$, by

$$\mathcal{P}_t(\rho, E) = \int_{\mathfrak{J}} \frac{\mathcal{P}_t(\tau, E) \cdot (\prod_{t'=1}^{t} \mu_{E,t'}(\tau)) dP_0}{\left[\int_{\mathfrak{J}} (\prod_{t'=1}^{t} \mu_{E,t'}(\tau')) dP_0\right]}$$

where P_0 is a probability measure (the initial distribution) on $\mathfrak{J}$. Roughly, ρ produces a mixture of distributions at each time t in which distribution $\mathcal{P}_t(\tau, E)$ occurs with density

$$\frac{(\prod_{t'=1}^{t} \mu_{E,t'}(\tau)) dP_0}{\left[\int_{\mathfrak{J}} (\prod_{t'=1}^{t} \mu_{E,t'}(\tau')) dP_0\right]}.$$

The name "reproductive" comes from the observation that the given densities result if each trajectory τ is thought of as increasing its proportion in a population of trajectories by reproducing at each time t according to its "payoff" $\mu_{E,t}(\tau)$. (When $\mathfrak{J}$ is discrete, the plan ρ reduces to assigning the probability

$$\sum_{\mathfrak{J}} \mathcal{P}_t(A, \tau, E) \cdot \frac{\prod_{t'=1}^{t} \mu_{E,t'}(\tau)}{[\sum_{\mathfrak{J}} \prod_{t'=1}^{t} \mu_{E,t'}(\tau')]}$$

to each $A \in \mathfrak{A}$. The reproductive plan can be defined for any case where $\mathfrak{A}$ and $\mathfrak{J}$ are both effectively defined and infinite by using the appropriate random functions with $\mathfrak{A}$ as the index set.) That the definition is meaningful is established by the

LEMMA. *Define $f_{E,t}: \mathfrak{J} \to \mathcal{P}$ by $f_{E,t}(\tau) = \mathcal{P}_t(\tau, E)$ and require that, for all $E \in \mathcal{E}$ and $t = 1, 2, \cdots$, $f_{E,t}$ is P_0-measurable with respect to $\mathfrak{J}$. (I.e., let the trajectories be so defined that the sets $\{\tau: \mathcal{P}_t(A_\alpha, \tau, E) \leqq P_\alpha$, all $A_\alpha \in \mathfrak{A}\}$ are measurable subsets of $\mathfrak{J}$ for all $E \in \mathcal{E}$, $t = 1, 2, \cdots$.) Then the expectation $\mathcal{P}_t(\rho, E)$ is defined and is a distribution for all $E \in \mathcal{E}$, $t = 1, 2, \cdots$.*

PROOF. $\prod_{t'=1}^{t}\mu_{E,t'}(\tau)$ is integrable with respect to P_0 so that

$$\int_{\mathfrak{J}}\left(\prod_{t'=1}^{t}\mu_{E,t'}(\tau)\right)dP_0$$

is defined and hence the function $g_{E,t}\colon \mathfrak{J}\to[0,1]$ defined by

$$g_{E,t}(\tau) = \frac{f_{E,t}(\tau)\prod_{t'=1}^{t}\mu_{E,t'}(\tau)}{\int_{\mathfrak{J}}(\prod_{t'=1}^{t}\mu_{E,t'}(\tau))dP_0}$$

is measurable and bounded (it is in fact a distribution over $\mathfrak{A}$). But then (the expectation)

$$\mathcal{P}_t(\rho, E) = \int_{\mathfrak{J}} g_{E,t}(\tau)dP_0$$

is defined. △

To show that ρ is $(\mathcal{E}, \mathfrak{J})$-good it must be established that the corresponding expected accumulation of losses $(\overset{*}{\mu}_{E,t}-\mu_{E,t}(\rho))$ is finite.

THEOREM. *Let M be any set of bounded continuous ranking functions μ_E indexed by $\mathcal{E}$ and such that for any $E\in\mathcal{E}$:*

(i) *there exists t for which $\overset{*}{\mu}_{E,t'}\geqq\overset{*}{\mu}_{E,t}>0$, for all $t'>t$. Let $\mathfrak{J}$ be any set of plans such that for each $E\in\mathcal{E}$:*

(i) *there exists a subset of good plans $\mathfrak{J}_E$ with $P_0(\mathfrak{J}_E)>0$,*

(ii) *the associated functions $f_{E,t}$ are P_0-measurable for $t=1, 2, \cdots$.*
Then the reproductive plan ρ is $(\mathcal{E}, \mathfrak{J})$-good.

PROOF. $\mu_{E,t}(\tau)=\mu_{E,t}(f_{E,t}(\tau))$ is a continuous function of a P_0-measurable function; hence the expected rank $\mu_{E,t}(\rho)$ of the mixture of distributions produced by ρ at time t is defined and is given by

$$\frac{\int_{\mathfrak{J}}\mu_{E,t}(\tau)(\prod_{t'=1}^{t}\mu_{E,t'}(\tau))dP_0}{\int_{\mathfrak{J}}(\prod_{t'=1}^{t}\mu_{E,t'}(\tau'))dP_0}.$$

Thus to show ρ is $(\mathcal{E}, \mathfrak{J})$-good it must be shown that

$$\lim_{T\to\infty}\sum_{t=1}^{T}\left[\frac{\int_{\mathfrak{J}}(\overset{*}{\mu}_{E,t}-\mu_{E,t}(\tau))(\prod_{t'=1}^{t}\mu_{E,t'}(\tau))dP_0}{\int_{\mathfrak{J}}(\prod_{t'=1}^{t}\mu_{E,t'}(\tau'))dP_0}\right] = N_E(\rho) < \infty.$$

Divide the numerator and denominator of the general term of the sum by $\prod_{t'=1}^{t} \overset{*}{\mu}_{E,t'}$ and apply the definition $(\overset{*}{\mu}_{E,t}-\mu_{E,t}(\tau))=\overset{*}{\mu}_{E,t}\epsilon_{E,t}(\tau)$, obtaining the new general term:

$$L_{E,t}(\tau) = \frac{\overset{*}{\mu}_{E,t}\epsilon_{E,t}(\tau)\prod_{t'=1}^{t}(1-\epsilon_{E,t'}(\tau))dP_0}{\int_{\mathfrak{J}}\prod_{t'=1}^{t}(1-\epsilon_{E,t'}(\tau'))dP_0}.$$

(When $\mathfrak{J}$ is finite, the theorem reduces to showing that the above sum is bounded above by

$$\frac{\overset{*}{\mu}_E}{\sum_{\mathfrak{J}}\prod_{t'=1}^{\infty}(1-\epsilon_{E,t'}(\tau'))}\cdot\sum_{t=1}^{\infty}\left(\epsilon_{E,t}(\tau)\prod_{t'=1}^{t}(1-\epsilon_{E,t'}(\tau))\right),$$

where $\overset{*}{\mu}_E$ is the highest possible rank in E. In this expression the first term is finite if there is a good plan for E and the second term is a convergent series (cf. Chapter IX of Knopp's *Infinite series*).) Now

$$(1-\epsilon_{E,t'}(\tau)) = \exp[\ln(1-\epsilon_{E,t'}(\tau))] \leqq \exp[-\epsilon_{E,t'}(\tau)].$$

Hence

$$L_{E,t}(\tau) \leqq \left[\frac{\overset{*}{\mu}_{E,t}}{\int_{\mathfrak{J}}\prod_{t'=1}^{t}(1-\epsilon_{E,t'}(\tau'))dP_0}\right]\cdot\left[\frac{\epsilon_{E,t}(\tau)dP_0}{\exp(\sum_{t'=1}^{t}\epsilon_{E,t'}(\tau))}\right].$$

Note that $\prod_{t'=1}^{t}(1-\epsilon_{E,t'}(\tau')) \geqq \prod_{t'=1}^{\infty}(1-\epsilon_{E,t'}(\tau'))$, all t. Thus

$$\lim_{T\to\infty}\sum_{t=1}^{T}\int_{\mathfrak{J}}L_{E,t}(\tau)dP_0 \leqq \left[\frac{\overset{*}{\mu}_E}{\int_{\mathfrak{J}}\prod_{t'=1}^{\infty}(1-\epsilon_{E,t'}(\tau))dP_0}\right]$$

$$\cdot\left[\lim_{T\to\infty}\sum_{t=1}^{T}\int_{\mathfrak{J}}\frac{\epsilon_{E,t}(\tau)dP_0}{\exp(\sum_{t'=1}^{t}\epsilon_{E,t'}(\tau))}\right]$$

$$\leqq \left[\frac{\overset{*}{\mu}_E}{\int_{\mathfrak{J}}\prod_{t'=1}^{\infty}(1-\epsilon_{E,t'}(\tau))dP_0}\right]$$

$$\cdot\left[\lim_{T\to\infty}\int_{\mathfrak{J}}\left(\sum_{t=1}^{T}\frac{\epsilon_{E,t}(\tau)}{\exp(\sum_{t'=1}^{t}\epsilon_{E,t'}(\tau))}\right)dP_0\right].$$

The terms $\epsilon_{E,t'}(\tau)$, $t'=1, \cdots, t$, in the sum occurring in the denominator of the second half of the product can be rearranged so that they are monotone decreasing. To show that the second half of the product is bounded above choose a monotone nonincreasing continuous $\beta_{T,E}(t)$ defined on $0 \leqq t \leqq T$ so that $\beta_{T,E}(t)=\epsilon_{E,t}(\tau)$ for $t=1, 2, \cdots, T$. Then

$$\int_{t'=1}^{t+1} \beta_{T,E}(t')dt' \leqq \sum_{t'=1}^{t} \epsilon_{E,t'}(\tau) \leqq \int_{t'=0}^{t} \beta_{T,E}(t')dt'.$$

Or, for any t and any $t-1 \leqq b \leqq t$,

$$\frac{\epsilon_{E,t}(\tau)}{\exp\left(\sum_{t'=1}^{t} \epsilon_{E,t'}(\tau)\right)} \leqq \frac{\epsilon_{E,t}(\tau)}{\exp\left(\int_{t'=1}^{t+1} \beta_{T,E}(t')dt'\right)} \leqq \frac{\beta_{T,E}(b)}{\exp\left(\int_{t'=1}^{b} \beta_{T,E}(t')dt'\right)}.$$

Hence, noting that

$$\frac{\epsilon_{E,1}(\tau)}{\exp(\epsilon_{E,1}(\tau))} \leqq 1,$$

$$\sum_{t=1}^{T} \frac{\epsilon_{E,t}(\tau)}{\exp\left(\sum_{t'=1}^{t} \epsilon_{E,t'}(\tau)\right)} \leqq 1 + \sum_{t=2}^{T} \frac{\epsilon_{E,t}(\tau)}{\exp\left(\sum_{t'=1}^{t} \epsilon_{E,t}(\tau)\right)} \leqq 1 + \int_{t=1}^{T} \frac{\beta_{T,E}(t)dt}{\exp\left(\int_{t'=1}^{t} \beta_{T,E}(t')dt'\right)}.$$

Substituting $X(t)$ for $\int_{t'=1}^{t} \beta_{T,E}(t')dt'$ yields for the integral on the right

$$\int_{X(1)}^{X(T)} \frac{dX}{e^X} = -\left.\frac{1}{e^X}\right|_{X(1)}^{X(T)} \leqq e^{-X(1)} = 1,$$

since $X(1)=\int_{t'=1}^{1} \beta_{T,E}(t')dt'=0$. But then

$$\lim_{T\to\infty} \int_{\mathfrak{Z}} \left(\sum_{t=1}^{T} \frac{\epsilon_{E,t}(\tau)}{\exp\left(\sum_{t'=1}^{t} \epsilon_{E,t'}(\tau)\right)} \right) dP_0 \leqq \lim_{T\to\infty} \int_{\mathfrak{Z}} (1+1)dP_0 = 2.$$

Therefore

$$\lim_{T\to\infty} \sum_{t=1}^{T} \int_{\mathfrak{I}} L_{E,t}(\tau)dP_0 \leqq \frac{2\overset{*}{\mu}_E}{\int_{\mathfrak{I}} (\prod_{t'=1}^{\infty}(1-\epsilon_{E,t'}(\tau')))dP_0}.$$

By hypothesis there exists a measurable subset $\mathfrak{I}_E$ of good plans for E (i.e., $\sum_{t'=1}^{\infty}\overset{*}{\mu}_{E,t}\epsilon_{E,t'}(\tau')$ converges on $\mathfrak{I}$) which implies that $\sum_{t'=1}^{\infty}\epsilon_{E,t'}(\tau)$ converges on $\mathfrak{I}_E$ because $\overset{*}{\mu}_{E,t'}$ is bounded below by $\overset{*}{\mu}_{E,t}>0$ for t' sufficiently large. Therefore $\prod_{t'=1}^{\infty}(1-\epsilon_{E,t'}(\tau'))$ converges (to a value greater than zero) on $\mathfrak{I}_E$ and the expectation

$$\int_{\mathfrak{I}}\left(\prod_{t'=1}^{\infty}(1-\epsilon_{E,t'}(\tau'))\right)dP_0$$

is greater than zero, which proves the theorem. △

The theorem can be generalized in several obvious ways (e.g., the μ_E need only be Baire functions and $\mathfrak{A}$, if given an appropriate structure, can be made uncountable) but these generalizations seem to have little to do with the intended interpretations.

UNIVERSITY OF MICHIGAN, ANN ARBOR, MICHIGAN 48104

GOAL-DIRECTED PATTERN RECOGNITION

by John H. Holland

Department of Communication Sciences
The University of Michigan

In theoretical investigations of pattern recognition it is common practice to represent the system-environment context minimally, usually by a partitioned set indicating the pattern categories. While this can be fruitful, as it has been for some kinds of character recognition, both thought and practice point up its limitations. It is particularly inadequate in goal-directed situations where the patterns to be recognized depend heavily on those features of the environment relevant to attainment of the system's goals. When the environment is unknown to the system in some essential aspects, such features must be discovered in the process of attempting to attain the goals. Different environments or different goals will determine different sets of critical features. That is, the critical features are defined implicitly by the system-environment context, rather than by being given explicitly. The object, then, of a goal-directed pattern recognition system is to discover, consistently recognize, and utilize critical features of any environmental situation within its competence. In terms more familiar to the pattern recognition field: the system, when confronted with any environment within its competence, must be able to classify each environmental state according to the goal-directed response to be made to it. Thus the system must develop a set of detectors appropriate to the particular environment confronting it while simultaneously developing a procedure for deciding the appropriate response to each input configuration presented by the detectors.

This paper is concerned with the problems entailed in constructing such systems and with ways of resolving these problems. Section 1 examines relevant aspects of Samuel's well-known work, with the intent of exhibiting some features of goal-directed pattern recognition in a familiar context. Then in Section 2 several critical problems are delineated. A general organizational plan which takes these problems into account is proposed in Section 3. Finally, in Section 4, this plan is examined in the guise of a sampling procedure, with suggestions as to how its performance should be analyzed.

1. Samuel's Checker Player as a Pattern Recognizer.

Samuel's checkerplayer[1] provides a familiar system employing goal-directed pattern recognition, though the pattern recognition aspect is usually not emphasized. The goal is clearcut: to win games. The input space in this case is the set B of possible checkerboard configurations. The candidates for critical features are functions θ_j from set B into the set of real numbers R. In Samuel's case, these functions are chosen initially. For the most part, they correspond to features an experienced checker player might look for, such as "pieces ahead", "penetration beyond the centerline", "control of the center", etc. For example, the function which measures "pieces ahead" assigns to each configuration $b \in B$ the difference in the number of pieces available to each side. It is significant for later discussion that these functions are actually defined by sequences of instructions for a digital computer and that the exact function defined does not seem to be overly critical. Once these functions have been chosen, they are combined in a weighted linear form, $V(b) = \sum_j a_{ij} \theta_j(b)$, in an attempt to assign a value

[1]Samuel, A. L., "Some Studies of Machine Learning Using the Game of Checkers," IBM J. Res. and Dev. 3, 211-229 (1959).

(relatedness to the goal of winning) to each board configuration $b \in B$.[2] Via a sophisticated "lookahead" procedure involving a model of the environment (including the opponent) an anticipated future configuration is assigned to each possible response (move). The <u>anticipated</u> configuration assigned by the model to each (immediate) response is then evaluated by V and the response associated with the highest-valued anticipated configuration is chosen. The use of a model as an integral part of the recognition process is a central concern of this paper.

The linear form $\sum_j a_{ij} \theta_j(b)$ used by Samuel is akin to the much-studied "adaptive threshold device": the functions θ_j correspond to the input functions of the threshold device, with the difference that their range is the reals rather than $\{0,1\}$; the threshold in effect varies from one set of move alternatives to another, being adjusted to select the highest-valued alternative in each set. It is important to note that the set of all possible arrays of values for the θ_j (i.e., the set $\pi_j R_j$, where R_j is the range of θ_j) constitutes the set of states of the checkerplayer's model of its environment. It has no way of distinguishing two boards b and b' if $[\theta_1(b), \ldots, \theta_m(b)] = [\theta_1(b'), \ldots, \theta_m(b')]$. Thus the adequacy of its model will depend in part upon whether or not distinctions which are critical to goal-attainment can in fact be made in the model.

---

[2] A later version ["Some Studies of Machine Learning Using the Game of Checkers. II-Recent Progress," <u>IBM J. Res. and Dev.</u> <u>11</u>, 601-617 (1967)] uses a more sophisticated technique for calculating V(b): a hierarchy of signature tables. Each signature table amounts to a multivalued truth table and the hierarchy results from using the values from one set of tables as arguments for another (higher-level) table. This section will concentrate on the earlier linear technique because of its closer relation to familiar techniques and because the value-estimator/model organization is more easily introduced on this basis. In Section 3, the method of signature tables provides a concrete (though restricted) matrix for the suggestions there.

The way in which Samuel adjusts and uses this linear form is not at all typical. The usual way is to provide an initial "training period" for adjusting the weights on the basis of information supplied by a referee capable of correctly classifying each test pattern presented. For the checkerplayer, on the other hand, a long sequence of patterns must be acted upon during the play of a game before a criterion (a "win" or a "loss") becomes available. The configurations underlying these patterns provide a great deal of relevant information, information which should be used in adjusting the weights. If it is not so used the checkerplayer encounters great difficulty in deciding which configurations were correctly, or incorrectly, handled in the course of play. It is here that the use of the linear form as a predictor opens up possibilities in just the right directions. A prediction about features of a configuration anticipated x moves hence can be checked as soon as x moves have elapsed. If the prediction is borne out, fine; if not, some adjustment of the linear form is called for. If, as suggested earlier, the prediction concerns the value of the expected configuration, then a heuristic for handling the adjustment of the linear form quickly presents itself. When the predicted value is off by an amount δ, the weight a_{ij} of function θ_j is modified to $a_{ij}-\Delta$, where $\Delta = c\delta a_{ij}\theta_j(b)$, b is the configuration at the time the prediction was made, and c is a constant of proportionality. (The algorithm actually used is more involved, particularly when the checkerplayer has an inexpert opponent, but the principle holds.) The result is a constant pressure for relevance and consistency in the prediction--relevance in the sense that the features recognized genuinely presage an opportunity (or difficulty), and consistency in the sense that the features are associated with the response appropriate to realizing (or avoiding) the prediction.[3] Irrelevant θ_j eventually acquire very low weights and hence are effectively eliminated from the linear form. Relevant θ_j acquire weights which, in conjunction with other θ_i, contribute to the choice of a goal-directed response. (Determination of an appropriate set of weights may be quite complicated since a given weighted θ_j may, without harm, take a moderate value in the "wrong direction" if at the same time other θ_i override it with large values in the "right direction"; this arrangement will be favored if there are other situations in which the same weighted θ_j proves to be the critical guide to an appropriate response.) At the end of each play of the game, the consistency of the value estimator will be checked against the outcome, thus tying the consistency requirement to the goal. The use of the linear form as a predictor enables the checkerplayer to assimilate the large amount of information it is receiving, continually sharpening its ability to recognize and use critical features.

2. Problems of Goal-Directed Pattern Recognition.

In Samuel's checkerplayer we see a particular case of an interesting organization for goal-directed pattern recognition: the combination of modifiable feature detectors with a "lookahead" model based on these detectors. Just how general and efficacious is this organization? Any attempt to answer this question leads almost inevitably to two subsidiary questions

---

[3]In other words: a set of detectors has sufficient relevance if all the distinctions critical to goal-attainment can be made within the set of model states $\pi_j R_j$ determined by the detectors. A transition function τ: O x $\pi_j R_j \rightarrow \pi_j R_j$, where O is the set of responses, yields a model based on the states $\pi_j R_j$; a model is consistent (with observation) if the observed state $[\theta_1(b'), \ldots, \theta_m(b)]$ after each response o is in fact the state $\tau[o, \theta_1(b), \ldots, \theta_m(b)]$ predicted by the transition function (where the environment goes from state b to state b' when response o is made). It is of course only necessary that the model be consistent for situations actually encountered.

which can be broadly phrased as follows: Can a system be supplied with procedures for generating extensive classes of feature detectors relevant to the environment confronting it? Are there reasonably general plans for constructing, modifying and using models on the basis of information supplied by the detectors?

Let us look more closely at the first of these questions. Samuel has already encountered a particular form of the problem presented by this question, designating it "the parameter problem." It arises when the set of detectors available to the system proves inadequate, in some circumstances, for prediction (i.e., the set has insufficient relevance). Since Samuel's detectors are defined by subroutines in a computer, new detectors can of course be generated by making random changes in the subroutines. However, by itself, this procedure is as little likely to prove effective as would an attempt to solve non-trivial problems by trying randomly modified programs one after another (cf. Friedberg).[4] Somehow, the detectors generated must be related to the needs of the system. Since the environment confronting the system will generally be unknown to it in several critical aspects, the system's needs and the detectors required must be discovered in the process of goal attainment. If the unknown aspects of the environment are at all complex, the detectors initially supplied can at best give the system a basis for discovering its further requirements. One fact uncovered by Samuel (and already alluded to) offers some hope. Fairly extensive alterations of the subroutines defining the basis functions θ_j did not much affect the performance of the checkerplayer. This indicates, at least in the case of the

[4] Friedberg, R. M., "A Learning Machine: Part 1," IBM J. Res. and Dev. 2, 2-13 (1958).

checkerplayer, that an exact hit is not necessary in order to find a new, relevant function. Each candidate function apparently belongs to a large class, each element of which is a usable alternative. What features of the checker-playing situation make this so; i.e., what are the general conditions underlying this phenomenon? How can they be exploited?

To properly treat these questions we must return to the second of the subsidiary questions posed earlier. In any realistic situation the flow of information (in the formal sense) is overwhelming. It is so overwhelming that it is not reasonable to think of producing a model which will predict the flow in all its detail. Still, for typical goal-oriented systems, most of the details of the flow are only marginally relevant. Herein lies the justification of critical feature detectors and at the same time the possibility of models relevant to goal attainment. The major purpose of the model, already exemplified in Samuel's case, is prediction of the relative value of the response open to the system at any point in time. If the detectors are adequate (i.e., not only relevant but sufficient) to the particular goal-environment combination, reliable predictions can be based on the much-reduced flow of information from the detectors. Given the adequacy of the detectors, the major problem is that of assuring consistency of predictions by continually checking the model against the information supplied by the detectors. The modifications required to improve consistency are straight-forward in Samuel's case because of the linear form. Since the contributions of the individual weighted functions (detectors) are linearly superposed, credit ("debit") for errors can be apportioned accordingly. (This is strictly analogous to the technique for sharpening the separation of pattern categories with an "adaptive threshold device.") However, even in those cases where the detectors adequately signal all critical information, it is questionable that they will behave in such a quasi-orthogonal fashion.

How is apportionment of credit to be made when the detectors enter into the predictor in more complicated fashion? This question is difficult enough but there is another as difficult. How is the system to determine whether or not the detectors it has are adequate?

The questions posed, to this point, are quite interdependent. The adequacy of a set of detectors for prediction cannot be determined in the absence of some assumptions about the predictor using them. The procedures for improving the predictor will depend upon the ease of generating and modifying detectors. The generation of relevant detectors will be contingent upon the way in which inadequacies in the extant set are discovered. And all depends upon the goals of the system and upon the environments the system may encounter.

3. A Value-Estimator/Model Procedure.

One thread through this labyrinth appears when the model-predictor relation is examined more closely.

In the usual case the system will be initially primed with a set of detectors and a provisional model of the environment based on the detectors. In effect these detectors constitute an initial representation of the environment, embodying relevant prior information. Each possible configuration of values for the set of detectors constitutes a state of the environment from the system's viewpoint. The initial model incorporates conjectures about changes of state which can be effected by system responses. In other words, the initial model specifies a transition function (conjectured and partial) for the environmental representation. By iterating this transition function it is possible to make predictions about future detector state-configurations. (If the representation is adequate and the model is both complete with respect to the responses and deterministic, then the state-configuration attained by a given sequence of responses can be determined by tracing the state transition diagram corresponding to the model, as for a finite automaton.) Given this basis, a major technique for response selection uses an estimation procedure to assign each detector state-configuration a measure of its relatedness to goals. This measure is intended to reflect the best that can be attained from each configuration under the most propitious selection of subsequent responses. In order to select a current response the model is used to look as far ahead as practicable, taking into account time available and reliability of the model. Typically this lookahead will be co-ordinated with the value estimator so that only responses or sequences of responses leading to highly-valued configurations are further explored. At the conclusion of the lookahead, that response is chosen which initiates the sequence leading to the highest-valued anticipated outcome.

The foregoing organization suggests an overall plan for tackling the problems outlined earlier. The two objectives remain: relevance of the detectors and consistency of the model and value-estimator.

The analogy which likens the model to a finite automaton suggests a procedure for detecting inadequacies in the set of detectors. In the case of a finite automaton the outcome of any input-state internal-state combination is always the same. In the case of the goal-directed system each detector state-configuration is supposed to designate a state of the environment. Thus, if the system is to have a strictly deterministic model of the environment, a given response should not, on successive occasions, produce different outcomes (detector state-configurations) from a given configuration. Any time such an aberration does occur an inadequacy of the set of detectors is indicated. Of course in many circumstances a strictly deterministic model will not be required; it will suffice if the outcomes of the response-state pair fall within some set of admissible variants. (E.g., if the environment is stochastic it may suffice that the variance of the outcomes not increase beyond some bound.) However, this just weakens the condition for aberration--the principle still holds.

Consider now the detectors as employed in a linear form. Assume that each time an aberration does occur the accompanying complete (unfiltered) input state is added to a collection of such instances. Any procedure which assigns a high value to instances in the collection, while assigning a low value to most other input states, would be a good candidate for a new detector. We see here some indication of why the exact definition of the detector is not critical, in part answering the earlier question based on Samuel's observations. If the set of situations to which the new detector assigns a high value contains a "low density" of non-aberrant situations, then there is a good opportunity for the other detectors to override the new detector on the non-aberrant situations (when it signals falsely). This suggests a plan for automatically generating relevant new detectors as the need arises, a plan closely related to recent observations in genetics concerning intra-chromosomal duplication.[5] As implemented here this would amount to adding to the set of detectors extensively modified versions of detectors which have high values at the time an aberration occurs. If the modification procedure is not too narrowly conceived, some of these new detectors will still take high values on the aberrant situations, while taking low values on other situations which the "parent" detector confused with the aberrant situations. They will enable the goal-directed system to enlarge the state-set of its model in a relevant fashion. Other new detectors, so generated, which are irrelevant will acquire near-zero weights, effectively eliminating them as earlier.

When the system uses a linear form, aberrations involving detectors

[5] Carnegie Institution of Washington Year Book 64, 331-333 (1965).

taking on relatively low values can be ignored since these detectors will have little effect on the value estimator. In effect, the detectors taking on low values are "inactive." However, the problem with the linear form is that an "active" (high-valued) detector is constrained to affect the value estimate strongly, and with the same sign, in all situations in which it is "active." A more general procedure would have the value estimator take different values for each distinct configuration of "active" and "inactive" detectors. But then the problem is one of determining which aberrations are important. The "active"-"inactive" dichotomy gives a hint as to how to proceed. Let the detectors be restricted to the range {0,1}, "inactive" or "active," and let the value estimation procedure be formulated as an independent subroutine which has the set of all possible detector value configurations as its domain. An aberration can then be handled as in the linear case, the active detectors at the time serving as the "parents" from which the new candidates are generated. Just as before, modifications of the detectors responsible for the aberrations have the best chance of distinguishing critical situations previously confused, thereby increasing the relevance of the set of detectors. At the same time, even for quite complex estimation functions, it will generally be easy to modify the function so that it assigns a lower (higher) value to a particular detector configuration.[6] Therefore (with a caveat yet to be discussed) the Δ-procedure earlier used with the linear form to improve its consistency can also be used with a more complex value estimator. Thus, there are procedures for increasing both relevance and consistency even in badly nonlinear situations.

4. The Procedure as a Sampling Plan.

As with other types of pattern recognition, the procedure just described

[6] See Samuel's discussion of the signature table procedure (ref. in footnote 2).

can be looked upon as a sampling procedure. On this view, the purpose of the detectors is to project the environment on a set of dimensions relevant to goal attainment. It is the resulting (reduced) sample space which makes feasible the construction of a goal-oriented model. One of the two main objectives of the system, then, is to produce a relevant model on the basis of samples drawn from the space. This it does by assigning to observed elements of the sample space the successor element produced by a chosen response. (Responses are chosen on the hypothesis that they lead toward goals.) Here an element new to most pattern recognition studies enters. The initially chosen dimensions may permit a model with aberrations--a model with inadmissible multiple consequences of a given response at a given sample point. Under such conditions the dimensioning must be changed. This requires sampling in a new and distinct sample space, the space of all admissible detectors (say, all subroutines of a certain class). The object of this second sampling plan is to use information obtained on the extant dimensions to discover "neighborhoods" which contain detectors likely to improve the dimensioning. (Our previous discussion indicates the liklihood of such "neighborhoods" vis-a-vis a model-predictor organization of the system.) These "neighborhoods" can then be exploited by an increased density of sampling therein. (The plan analogous to intrachromosomal duplication does just this--responding to a dimensioning "crisis" by acting upon just those dimensions relevant to the crisis.) In so-doing the second plan seeks to resolve the counterpart of Samuel's parameter problem by locating new, relevant detectors.

Simultaneously with these two sampling procedures a third is carried out, aimed at building up a consistent value estimator. This constitutes the second major objective of the system. At each stage a candidate estimator is tested against the model for the consistency of its predictions. When a prediction is not verified the object is to select a new estimator which preserves the good features of the failed estimator while correcting for its inadequacy. This objective will be facilitated if credit for the error can be apportioned to the parts of the estimator, but this requires a suitable decomposition theory for the estimators. With a linear estimator there is a direct decomposition in terms of weighted detector functions. No such obvious decomposition exists for more sophisticated estimators, but there is a useful approach stemming from mathematical genetics. It is based on the assignment of "average excesses" as a measure of the average contribution of a given detector to the overall value estimate.[7] One can thus use the average excesses to apportion credit for errors and proceed as in the linear case. (There is the additional advantage of an interesting mathematical theory related to the rate of convergence of this technique.)[8]

Given enough time the system will acquire a relevant set of detectors and a consistent estimator. At that point a response can be rated immediately, without any further lookahead, by looking at the estimated value of the detector configuration it will lead to (according to the model). The effect is that of responding directly to the critical features detected.

The fundamental idea underlying all phases of the sampling plan's operation can be expressed as a testing of schema. In this context a schema results when a part of some structure (a detector, an estimator) is deleted and the resulting vacancy is treated as a slot in which any other structure satisfying the interface conditions can be substituted. If the structure is a program, then any subroutine with the proper entries and exits would be a candidate for filling the vacancy. The favored schema preserve critical pieces of goal-relevant information accumulated in past tests; the substitution instances

---

[7]Kimura, M., "On the Change of Population Fitness by Natural Selection." _Heredity_ _12_ 145-167 (1958).

[8]_Idem._

are sampled on the basis of a probability distribution skewed according to past results. Apportionment of credit provides one way of determining what schema should be favored and what probabilities should be assigned to different instances. The distributions so determined play a role much like that of a simplicity ordering in inductive inference. It is important to note that each test of a particular structure, if properly exploited, constitutes a (simultaneous) test of all the schema which can be formed from that structure. This inherent parallelism provides a greatly increased sampling rate in the space of schema, with all the advantages attendant thereon.

- * -

The fundamental problem of goal-directed pattern recognition is to discover plans for generating and testing schema, plans which are demonstrabl good over a broad spectrum of environments. In this paper we have looked at a general organization for accomplishing this end and some operators for implementing the organization.[9] On one hand the organization generalizes a successful, empirically-tested approach; on the other, it makes contact with some important aspects of genetics and mathematical genetics. We have looked at only one genetic-related operator but there are many candidates. Crossover, inversion, operon-operator concatenations, and others, all have meaning here, and some versions offer demonstrable improvements over techniques currently used to attain the same ends (see Bagley[10] and Rosenberg[11]). To date, these operators and the attendant theory have been little studied or exploited in a

---

[9]A more mathematical discussion of this organization can be found in Holland, J. H., "Nonlinear Environments Permitting Efficient Adaption" in Computer and Information Sciences - II (ed. J. T. Tou) Academic Press (1967).

[10]Bagley, J. D., The Behavior of Adaptive Systems Which Employ Genetic and Correlation Algorithms. U. of Michigan Ph.D. Dissertation (1967).

[11]Rosenberg, R. S., Simulation of Genetic Populations with Biochemical Properties. U. of Michigan Ph.D. Dissertation (1967).

goal-directed sampling format. In my opinion a greater effort would be amply repaid, both by an increased sophistication in sampling plans for goal-directed pattern recognition and by an enlarged understanding of the action of genetic operators.

Acknowledgements

I would like to thank Professor Donald F. Stanat for a careful reading of the first draft of this paper and for many helpful suggestions. I would also like to thank the Office of Naval Research for its continued financial support (currently under Contract No. N00014-67-A-0181-0011) in areas relevant to the present paper.

ROBUST ALGORITHMS FOR ADAPTATION SET IN A GENERAL FORMAL FRAMEWORK

John H. Holland
The University of Michigan
Ann Arbor, Michigan

Summary

This paper presents a general formalism for defining problems of adaptation. Within this formalism a class of algorithms, the sequential reproductive plans, is defined. It is then proved that sequential reproductive plans are robust in the sense that, in environments with a well-defined measure of payoff (utility, performance), any such plan will acquire payoff at a rate which asymptotically approaches the optimum.

Questions about adaptation figure prominantly in studies of a wide variety of systems, both natural and artificial. Unfortunately in most cases of interest the questions are difficult to formulate precisely, let alone with generality. The attendant confusion sometimes grows to the extent that adaptation is considered a questionable way of extending system capability, and even to the extent that the existence of adaptation is disputed (except in the trivial sense of preserving the best in a sequence of random trials). These difficulties can be resolved if we can find a formal setting providing a uniform treatment of adaptation in its different guises. Although the general study of adaptation ultimately sets its own unique requirements upon formalism, we can draw upon a good deal of relevant experience in translating intuitive questions about computing and system organization into a formal setting. We can also take warning from this experience: Without considerable care the original questions often undergo a metamorphosis which leaves them only distantly related to their origins; the resulting answers, though precise and sometimes elegant, have little relevance in the areas of intended application.

The Formal Framework

This paper will approach a relevant formal framework by concentrating on the question: How can problems requiring (or at least encouraging) adaptation be distinguished?

No matter what the guise, we find that adaptation involves a progressive change of structure or organization in response to the environment or milieu of the system. Formally we can represent this process by designating a class of possible or admissible structures $\mathcal{A}$ and a set of operators $\Omega = \{\omega: \mathcal{A} \to \mathcal{A}\}$ which can be used to modify the structures. The following table lists typical structures and operators for some fields of interest:

Field	Structures	Operators
Genetics	chromosomes	mutation, recombination, etc.
Economic Planning	mixes of goods	production activities
Control	policies	Bayes's rule, successive approximation, etc.
Physiological Psychology	cell assemblies	synapse modification
Game Theory	strategies	rules for iterative approximation of minimax strategy
Artificial Intelligence	programs	"learning" rules

In all cases of interest, adaptation is a response to an initial uncertainty about the system's environment. That is, one or more relevant characteristics of the environment are unknown to the system, alternatives being possible; the set of all possible combinations of alternatives for the unknown characteristics constitutes the range of environments in which the system must be able to act if it is to exhibit adaptation. Formally we represent this range by a class $\mathcal{E}$ of possible environments. The process of adaptation amounts to making successive selections from Ω under the influence of information received from the environment $E \in \mathcal{E}$, actually confronting the system. Different algorithms or plans for making the selections amount to different methods of adaptation. To make this notion precise let S_E designate the possible states of environment $E \in \mathcal{E}$, and let $S_E(t)$ designate the state of that environment at time t. Let the structure of the adapting system at time t be $\mathcal{A}(t) \in \mathcal{A}$. Then the information received by the plan will be given by the function

$$\iota: S_E \times \mathcal{A} \to I,$$

where I is the set of possible inputs to the system, and the particular information received by the plan at time t is

$$I(t) = \iota(S_E(t), \mathcal{A}(t)).$$

By requiring that $\mathcal{A}(t)$ summarize whatever past history is to be available to the adaptive plan at time t, we can represent the plan formally by the two argument function

$$\tau: I \times \mathcal{A} \to \Omega.$$

If

$$\omega_t = \tau(I(t), \mathcal{A}(t))$$

designates the particular operator selected by plan τ at time t then

$$\mathcal{A}(t+1) = \omega_t(\mathcal{A}(t)) = [\tau(I(t), \mathcal{A}(t))](\mathcal{A}(t))$$

gives the modified structure produced by the plan. Within this framework the general objective will be to compare various adaptive plans, i.e., functions τ of the type just defined, either as hypotheses about natural phenomena or as algorithms for artificial systems. To carry out this comparison we must have in mind some set of feasible or possible plans $\mathcal{T}$, either the set of all possible functions from $I \times \mathcal{A}$ into Ω or some constrained subset of this set. We must also have some criterion of comparison χ. Bringing all of this together, a problem of adaptation will be said to be precisely posed when the triple $(\mathcal{E}, \mathcal{T}, \chi)$ is specified according to the above definitions.

Reproductive Plans

The $(\mathcal{E}, \mathcal{T}, \chi)$ framework suggests a search for adaptive plans which are robust in the sense that they meet some fairly stringent criterion χ over a wide range of conditions $\mathcal{E}$. Most of the rest of this paper will be devoted to proving robustness for a class of plans, the reproductive plans, which will be characterized as follows:

Reproductive plans will be restricted to environments in which some well defined measure of payoff (utility, performance) μ can be assigned to each structure in $\mathcal{A}$. That is, for each $E \in \mathcal{E}$, there must be a function

$$\mu_E: \mathcal{A} \to (r_1, r_2)$$

where (r_1, r_2) is the set of reals between r_1 and r_2, and the payoff $\mu_E(A)$ rates A's performance in the environment $E \in \mathcal{E}$. Reproductive plans will act only on information about payoff; more precisely,

$$I = (r_1, r_2)$$

and

$$I(t) = \mu_E(\mathcal{A}(t)).$$

(It should be clear that reproductive plans set a non-trivial lower bound on the performance of plans employing additional information, since at worst plans can ignore additional information and act only on payoff). The paradigm upon which reproductive plans are based, and the origin of the name, emerges when we think of the plan as being employed or represented by a set of $N(t)$ individuals at time t. If each of these individuals produces offspring in proportion to the payoff received by the corresponding plan then there will be

$$N(t+1) = \mu_E(\mathcal{A}_{\tau,E}(t)) \cdot N(t)$$

individuals at time t+1, assuming that in environment E plan τ selects structure $\mathcal{A}_{\tau,E}(t)$ at time t. Assuming $N(0) = 1$, there will be

$$N(t) = \Pi_{t'=1}^{t} \mu_E(\mathcal{A}_{\tau,E}(t'))$$

individuals representing the plan at time t. If we think now of several plans $(\tau_1, \ldots, \tau_H)$ operating simultaneously there will be a total of

$$N(t) = \sum_{j=1}^{H} \Pi_{t'=1}^{t} \mu_E(\mathcal{A}_{\tau_j,E}(t'))$$

individuals at time t and plan τ_j will be employed by a fraction

$$P_{\tau_j,E}(t) =$$

$$[\Pi_{t'=1}^{t} \mu_E(\mathcal{A}_{\tau_j,E}(t'))]/[\sum_{i=1}^{H} \Pi_{t'=1}^{t} \mu_E(\mathcal{A}_{\tau_i,E}(t'))]$$

of the individuals. For given t we can think of the numbers $P_{\tau_j,E}(t)$ as a distribution over the set $\{\tau_1, \ldots, \tau_H\}$. With some slight adjustments, it is this distribution determined by the reproductive paradigm (the product of payoffs) which will be used to determine a kind of mixed plan (cf. mixed strategy) which has the desired robustness. Precisely, the operation of a <u>sequential reproductive plan</u> is defined relative to any set $\mathcal{T}$ of adaptive plans as follows: At each time t the reproductive plan probabilistically chooses one plan $\tau \in \mathcal{T}$ to generate the structure to be tested at that time. If τ has already been chosen $N(\tau,t)$ times prior to t, then it will be used to generate $\mathcal{A}_{\tau,E}(N(\tau,t)+1)$, the $(N(\tau,t)+1)$th structure in the sequence of structures $\langle \mathcal{A}_{\tau,E}(1), \mathcal{A}_{\tau,E}(2), \ldots \rangle$ τ would ordinarily generate in E. Let $\mathcal{T}(t) = \{\tau_1, \tau_2, \ldots, \tau_h\}$ be the set of plans tried at least once prior to time t and let t_j be the time τ_j was first tried by the reproductive plan, with $t_j < t_{j+1}$ for all j. Then the probability of retrying the jth plan in $\mathcal{T}(t)$ when the environment is $E \in \mathcal{E}$ is given by

$$P_{j,t} = (1-c_{h+1}) \frac{c_j \Pi_{t'=t_j}^{t} (\mu_{E,t'}(\tau_j)+k_1)^{k_2}}{\sum_{h'=1}^{h} c_{h'} \Pi_{t'=t_{h'}}^{t} (\mu_{E,t'}(\tau_{h'})+k_1)^{k_2}},$$

where k_1 and k_2 are parameters which can be chosen arbitrarily within the limits $-r_1 < k_1 < \infty$ and $0 < k_2 < \infty$, c_j belongs to an infinite series of the form $\langle j^{-(1+\delta)} \rangle$, $\delta > 0$, and $\mu_{E,t}(\tau)$ is defined by

$$\mu_{E,t}(\tau) = \mu_E(\mathcal{A}_{\tau,E}(t))$$

if τ used at time t,

$$= \mu^*_{E,t} = \underset{\tau' \in \mathcal{T}}{\text{lub}} \{\mu_E(\mathcal{A}_{\tau',E}(t))\}$$

if τ not used at time t,

The probability of trying a completely new plan $\tau_{h+1} \in \mathcal{T} - \mathcal{T}(t)$ to generate the structure to be tested at time t is given by

$$P_{h+1,t} = c_{N(t)+1}$$

where $N(t)$ = the number of elements in $\mathcal{T}(t)$; τ_{h+1} is then chosen from $\mathcal{T} - \mathcal{T}(t)$ according to a fixed "initial" probability distribution over $\mathcal{T}$, $P_{\mathcal{T}}(0)$ (e.g., each element of $\mathcal{T}$ is treated as equally likely). It should be clear that, at each time t,

$$\sum_{j=1}^{N(t)+1} P_{j,t} = 1$$

and hence the set $\{P_{j,t}, j=1, \ldots, N(t)+1\}$ determines a distribution over $\mathcal{T}$. This is the distribution alluded to in the above paragraph.

The sequence of distributions over $\mathcal{T}$

$$\langle \{P_{j,t}, j=1, \ldots, N(t)+1\}, t=1,2,\ldots \rangle$$

defines a sequence of random variables corresponding to the succession of choices to be made by the sequential reproductive plan. Thus, the definition of the distributions in terms of the received payoffs $\mu_{E,t}(\tau)$ gives a complete stochastic definition of the plan.

<u>Robustness of Reproductive Plans</u>

Informally, the proof that follows shows that any sequential reproductive plan is almost "universally" robust, being proof against false peaks, non-linearities, etc. More carefully, in any environment $E \in \mathcal{E}$ with an arbitrary associated payoff function $\mu_E: \mathcal{A} \rightarrow (r_1, r_2)$ and relative to any "reasonable" set of feasible plans, a sequential reproductive plan will:

(1) suffer only finite cumulative expected losses when $\mathcal{T}$ is finite (e.g., when $\mathcal{T}$ is the set of n pure strategies for the N-armed bandit problem);
(2) symptotically approach the optimal average rate of payoff when $\mathcal{T}$ is infinite (e.g., when $\mathcal{T}$ is the set of all economic plans for a von Neumann technology).

The following definitions enable a concise, rigorous statement of the theorem:

An <u>arena</u> is defined as any pair $(\mathcal{T}, \mathcal{E})$ such that $(\forall E \in \mathcal{E})(\exists \mathcal{T}_E \subset \mathcal{T}) [\mathcal{T}_E$ has a probability greater than 0 under $P_{\mathcal{T}}(0)$ and $\underset{\tau \in \mathcal{T}_E}{\text{glb}} \{\mu_{E,t}(\tau) \geq \mu^*_{E,t} - \lambda_{E,t}$ where $\sum_{t=1}^{\infty} \lambda_{E,t} = N_E < \infty]$.

(The condition for an arena will automatically be satisfied if for each $E \in \mathcal{E}$ there is a "best" plan within $\mathcal{T}$ with an assigned initial probability greater than zero).

The <u>loss</u> associated with a plan τ at time t will be defined to be

$$L_{E,t}(\tau) = \mu^*_{E,t} - \mu_{E,t}(\tau)$$

(Note that this means that τ suffers zero loss if it is not tried at time t, since $\mu_{E,t}(\tau) = \mu^*_{E,t}$ then).

$\Psi_{E,t}(\rho)$ = the distribution assigned to $\mathcal{T}$ by the sequential reproductive plan ρ in environment E at time t.

$U_{E,t}(\rho)$ = the expected payoff associated with ρ at time t under $\tau_{E,t}(\rho)$. (Note that $\mu^*_{E,t}-U_{E,t}(\rho)$ is the expected loss associated with ρ at time t).

Theorem: Given an arena $(\mathcal{T}, \mathcal{E})$, for each $E \in \mathcal{E}$ a sequential reproductive plan ρ will with probability 1 satisfy the following criteria:

(1) $\lim_{T\to\infty} \sum_{t=1}^{T}(\mu^*_{E,t}-U_{E,t}(\rho)) < \infty$ if $\mathcal{T}$ is finite;

(2) $\lim_{T\to\infty} [\sum_{t=1}^{T}U_{E,t}(\rho)/\sum_{t=1}^{T}\mu^*_{E,t}] = 1$ if $\mathcal{T}$ is infinite.

Proof:

(Since the proof is for arbitrary $E \in \mathcal{E}$ the subscript E will be dropped throughout, except when dependence upon the environment is being particularly emphasized).

Abbreviating $\mu_{Et}(\tau_j)$ by u_{jt} and using the definition of expected payoff we get

$$U_t(\rho) = \sum_{j=1}^{N(t)}\mu_{jt}P_{jt} + c_{N(t)+1}\bar{X}_{\mathcal{T}-\mathcal{T}(t)},$$

where $\bar{X}_{\mathcal{T}-\mathcal{T}(t)}$ is the expected payoff when $\tau_{N(t)+1}$ is drawn from $\mathcal{T}-\mathcal{T}(t)$ according to $P_{\mathcal{T}}(0)$.

It will be convenient to rewrite $\sum_{t=1}^{T}U_t(\rho)/\sum_{t=1}^{T}\mu^*_t$ in terms of losses using the definition $P'_{jt} = P_{jt}/(1-c_{N(t)+1})$:

$$\frac{\sum_{t=1}^{T}U_t(\rho)}{\sum_{t=1}^{T}\mu^*_t} = 1 - \frac{\sum_{t=1}^{T}\mu^*_t - \sum_{t=1}^{T}U_t(\rho)}{\sum_{t=1}^{T}\mu^*_t}$$

$$= 1 - \frac{\sum_{t=1}^{T}(\mu^*_t - \sum_{j=1}^{N(t)}\mu_{jt}P'_{jt})}{\sum_{t=1}^{T}\mu^*_t} - \frac{\sum_{t=1}^{T}c_{N(t)+1}\mu^*_t}{\sum_{t=1}^{T}\mu^*_t}$$

$$+ \frac{\sum_{t=1}^{T}c_{N(t)+1}(\mu^*_t - \sum_{j=1}^{N(t)}\mu_{jt}P'_{jt})}{\sum_{t=1}^{T}\mu^*_t}$$

$$+ \frac{\sum_{t=1}^{T}c_{N(t)+1}\bar{X}_{\mathcal{T}-\mathcal{T}(t)}}{\sum_{t=1}^{T}\mu^*_t}$$

Label the last four terms on the right side L_{1T}, L_{2T}, L_{3T} and L_{4T} respectively, so that

$$\frac{\sum_{t=1}^{T}U_t(\rho)}{\sum_{t=1}^{T}\mu^*_t} = 1 - L_{1T} - L_{2T} + L_{3T} + L_{4T}.$$

It is clear that L_{3T} is always positive and that, for all T, $L_{4T} < L_{2T}$. Thus, if it can be shown that $\lim_{T\to\infty} L_{1T} = 0$ and $\lim_{T\to\infty} L_{2T} = 0$, the first part of the theorem will have been proved. It should also be clear that, when $\mathcal{T}$ is finite, there will with probability 1 be a time T_* after which every plan in $\mathcal{T}$ has been tried at least once. For $t > T_*$ losses will all arise from retrials of $\mathcal{T}(t) = \mathcal{T}(T_*) = \mathcal{T}$, whence the second part of the theorem can be proved by showing that $\lim_{T\to\infty} L_{1T} = 0$ because its numerator remains finite in the limit. The proof will proceed accordingly.

To show that $\lim_{T\to\infty} L_{2T} = 0$, note that with probability 1 there will be a time t_1 such that $c_{N(t_1)} < \frac{\epsilon}{2}$ since $\langle c_i \rangle$ is a monotone decreasing sequence. There also exists t_ϵ such that

$$\sum_{t=1}^{t_1}c_{N(t)}\mu^*_t/\sum_{t=1}^{t_\epsilon}\mu^*_t < \frac{\epsilon}{2}$$

because for all t $\mu^*_t > 0$, which means that the denominator can be increased without limit, while the numerator is finite for fixed t_1. For $T > t_\epsilon$

$$\frac{\sum_{t=1}^{T}c_{N(t)}\mu^*_t}{\sum_{t=1}^{T}\mu^*_t} = \frac{\sum_{t=1}^{t_1}c_{N(t)}\mu^*_t + \sum_{t=t_1}^{T}c_{N(t)}\mu^*_t}{\sum_{t=1}^{T}\mu^*_t}$$

$$< \frac{\epsilon}{2} + \frac{\frac{\epsilon}{2}\sum_{t=t_1}^{T}\mu^*_t}{\sum_{t=1}^{T}\mu^*_t} < \epsilon$$

Thus, given any $\epsilon > 0$, there exists t_ϵ such that, for $T > t_\epsilon$, $L_{2T} < \epsilon$. Hence $\lim_{T\to\infty} L_{2T} = 0$

To show that the numerator of L_{1T} remains finite in the limit, three definitions are helpful:

Define $c_{E,t}(\tau) = 1 - \dfrac{\mu_{E,t}(\tau) + k_1}{\mu^*_{E,t} + k_1}$

Define $\mathcal{T}_{E,r} = \{\tau \ni K_{r-1} \geq \Pi_{t=1}^{\infty}(1-c_{E,t}(\tau)) > K_r\}$

where r=0,1,2,..., the K_r are monotone decreasing to zero, $K_0 = 1$ and, for all r, $K_r > 0$.

Define $P_{Er} = P_{\mathcal{T}}(\mathcal{T}_{Er},0)$, the probability assigned to $\mathcal{T}_{Er} \subset \mathcal{T}$ by $P_{\mathcal{T}}(0)$.

Using the definition of arena we can show that there must be an integer q such that $P_{Eq} > 0$,

$$\mu^*_t - \mu_t(\tau) = (\mu^*_t + k_1)c_t(\tau)$$
$$\leq \lambda_t \text{ if } \tau \in \mathcal{T}_E$$

That is, $\sum_{t=1}^{\infty}(\mu^*_t+k_1)c_t(\tau) \leq \sum_{t=1}^{\infty}\lambda_t = N_E < \infty$

Therefore, $\sum_{t=1}^{\infty}c_t(\tau)$ converges for $\tau \in \mathcal{T}_E$ since $(\mu^*_t+k_1) > 0$ for all t.

But then $\Pi_{t=1}^{\infty}(1-c_t(\tau))$ converges (to a value greater than zero) and hence there must be a subset $\mathcal{T}_{Eq}$ as required because $\mathcal{T}_E$ has a probability greater than 0 and therefore one of the subsets $\mathcal{T}_{Er}$ must also have a probability greater than zero.

Substituting for the P'_{jt} in L_{1T} gives the following expression for the numerator:

$$L'_{1T} \overset{df.}{=} \sum_{t=1}^{T} L''_{1t}$$

$$= \sum_{t=1}^{T}\sum_{j=1}^{N(t)} (\mu_t^* - \mu_{jt}) \frac{c_j \Pi_{t'=t_j}^{t} (\mu_{jt'} + k_1)^{k_2}}{\sum_{h'=1}^{h} c_{h'} \Pi_{t'=t_{h'}}^{t} (\mu_{h't'} + k_1)^{k_2}}$$

abbreviate $\Pi_{t'=t_j}^{t}(\mu_{jt'}+k_1)^{k_2}$ by Π_{jt}. Let L_{1it} be the expected loss at time t assuming an element of $\mathcal{T}_{Eq}$ is first encountered at time t_i.

$$L_{1it} < \sum_{j=1}^{N(t)} (\mu_t^* - \mu_{jt}) \frac{c_j \Pi_{jt}}{c_i K_q + \sum_{h' \neq i} c_{h'} \Pi_{h't}} \overset{df.}{=} {}^{b}L_{1it}$$

since, by definition, $\tau_i \in \mathcal{T}_{Eq} \Rightarrow \Pi_{it} > K_q$. On this assumption of first encounter at time t_i, the cumulative expected loss is bounded by

$$\left(\sum_{t=t_i+1}^{\infty} {}^{b}L_{1it}\right) + (r_2 - r_1)t_i$$

since the loss associated with any given trial cannot exceed $r_2 - r_1$. Noting that the t_i are random variables, the overall cumulative expected loss can be bounded by

$$\sum_{i=1}^{\infty} (1-P_{Eq})^{i-1} P_{Eq} \left[\left(\sum_{t=t_i+1}^{\infty} {}^{b}L_{1it}\right) + (r_2 - r_1)\bar{t}_i\right]$$

$$= \sum_{i=1}^{\infty} (1-P_{Eq})^{i-1} P_{Eq} \sum_{t=t_i+1}^{\infty} {}^{b}L_{1it}$$

$$+ (r_2 - r_1) \sum_{i=1}^{\infty} (1-P_{Eq})^{i-1} P_{Eq} \bar{t}_i$$

where $\bar{t}_i$ is the expected value of t_i. To show that the numerator of L_{1t} remains finite, it is necessary to show that the last two terms are finite:

Concerning $(r_2 - r_1) \sum_{i=1}^{\infty} (1-P_{Eq})^{i-1} P_{Eq} \bar{t}_i$:

$\bar{t}_i = \sum_{x=1}^{\infty} P_i(x)x$, where $P_i(x)$ is the probability that $t_i = x$

$= \sum_{x=1}^{\infty} \sum_{(h_2,\ldots,h_i) \ni \sum_{\ell=2}^{i} h_\ell = x} P(h_2,\ldots,h_i)x$, where $P(h_2,\ldots,h_i)$ is the joint probability that $t_\ell - t_{\ell-1} = h_\ell, \ell = 2,\ldots,i$.

$= \sum_{\ell=2}^{i} \sum_{h=1}^{\infty} P_\ell(h)h$

$= \sum_{\ell=2}^{i} (\frac{1}{c_\ell})$, because the interval $(t_{\ell-1}, t_\ell)$ consists of a sequence of independent Bernoulli trials with probability c_ℓ

$< i(\frac{1}{c_i})$, since c_i decreases monotonely to zero

$< i^{2+\delta}$, substituting for c_i.

accordingly,

$$(r_2 - r_1) \sum_{i=1}^{\infty} (1-P_{Eq})^{i-1} P_{Eq} \bar{t}_i$$

$$< (r_2 - r_1) \sum_{i=1}^{\infty} (1-P_{Eq})^{i-1} P_{Eq} (i)^{2+\delta}$$

which is convergent by Cauchy's ratio test.

Concerning $\sum_{i=1}^{\infty} (1-P_{Eq})^{i-1} P_{Eq} \sum_{t=t_i+1}^{\infty} {}^{b}L_{1it}$:

$$< \sum_{i=1}^{\infty} (1-P_{Eq})^{i-1} P_{Eq} \sum_{t=1}^{\infty} \sum_{j=1}^{N(t)} (\mu_t^* - \mu_{jt})$$

$$\cdot \frac{c_j \Pi_{jt}}{c_i K_q + \sum_{h' \neq i} c_{h'} \Pi_{h't}}$$

Dividing the numerator and denominator by $\Pi_{t'=1}^{t}(\mu_{t'}^* + k_1)^{k_2}$ and using the definition of $\epsilon_{E,t}(\tau_j)$, which will be abbreviated to ϵ_{jt}, gives

$$< \sum_{i=1}^{\infty} (1-P_{Eq})^{i-1} P_{Eq} \sum_{t=1}^{\infty} \sum_{j=1}^{N(t)} \mu_t^* \epsilon_{jt}$$

$$\cdot \frac{c_j \Pi_{t'=1}^{t} (1-\epsilon_{jt'})^{k_2}}{c_i K_q + \sum_{h' \neq i} c_{h'} \Pi_{t'=1}^{t} (1-\epsilon_{h't'})^{k_2}}$$

$$< \frac{P_{Eq}}{K_q} \sum_{t=1}^{\infty} \mu_t^* \sum_{j=1}^{N(t)} \epsilon_{jt} c_j \Pi_{t'=1}^{t} (1-\epsilon_{jt'})^{k_2} \sum_{i=1}^{\infty} \frac{(1-P_{Eq})^{i-1}}{c_i}$$

Again by Cauchy's ratio test $\sum_{i=1}^{\infty} \frac{(1-P_{Eq})^{i-1}}{c_i}$ is convergent, say to the value $B(\delta)$ when $c_i = (i)^{-(1+\delta)}$.

From this it follows that numerator of L_{1T} remains finite if $\sum_{j=1}^{N(T)} c_j \sum_{t=1}^{T} \epsilon_{jt} \Pi_{t'=1}^{t} (1-\epsilon_{jt'})^{k_2}$ is bounded above for all T. To determine such a bound note first that,

$(1-\epsilon)^k = \exp[\ln(1-\epsilon)^k] = \exp[k\ln(1-\epsilon)] \leq e^{-k\epsilon}$.

Hence $\epsilon_{jt} \Pi_{t'=1}^{t} (1-\epsilon_{jt'})^{k_2} \leq \epsilon_{jt} / \exp[k_2 \sum_{t'=1}^{t} \epsilon_{jt'}]$

For any given t the terms in $\sum_{t'=1}^{t} \epsilon_{jt'}$ can be rearranged so that they are monotone decreasing; it is then easy to choose a monotone nonincreasing continuous $\beta_j(t)$ defined on $0 \leq t \leq T$ so that $\beta_j(t) = \epsilon_{jt}$ for $t = 1,2,\ldots,T$. Then

$$\int_{t'=1}^{t+1} \beta_j(t')dt' \leq \sum_{t'=1}^{t} \epsilon_{jt'} \leq \int_{t'=0}^{t} \beta_j(t')dt'.$$

Or, for any t and any $t-1 \leq b \leq t$,

$$\frac{c_{jt}}{\exp[k_2\sum_{t'=1}^{t} c_{jt'}]} \le \frac{c_{jt}}{\exp[k_2\int_{t'=1}^{t+1}\beta_j(t')dt']}$$

$$\le \frac{\beta_j(b)}{\exp[k_2\int_{t'=1}^{b}\beta_j(t')dt']}$$

Hence, noting that $c_{j1}/\exp[c_{j1}] \le 1$,

$$\sum_{t=1}^{T} \frac{c_{jt}}{\exp[k_2\sum_{t'=1}^{t} c_{jt'}]} \le 1 + \frac{1}{k_2}\sum_{t=2}^{T} \frac{k_2 c_{jt}}{\exp[k_2\sum_{t'=1}^{t} c_{jt'}]}$$

$$\le 1+\frac{1}{k_2}\int_1^T \frac{k_2\beta_j(t)dt}{\exp[\int_1^t k_2\beta_j(t')dt']}$$

Substituting $X_j(t)$ for $\int_1^t k_2\beta_j(t')dt'$ in the last integral yields

$$\int_{X(1)}^{X(T)} \frac{dX}{e^X} = -\frac{1}{e^X}\Big|_{X(1)}^{X(T)} \le 1 \text{ for all } T.$$

Hence $\sum_{t=1}^{T} c_{jt}\Pi_{t'=1}^{t}(1-c_{jt'})^{k_2} \le 1+\frac{1}{k_2}$, for all T,

and

$$\sum_{j=1}^{N(T)} c_j \sum_{t=1}^{T} c_{jt}\Pi_{t'=1}^{t} (1-c_{jt'})^{k_2} \le \sum_{j=1}^{N(T)} c_j(1+\frac{1}{k_2})$$

$$\le (1+\frac{1}{k_2})\sum_{j=1}^{\infty} c_j, \text{ for all } T,$$

which demonstrates the required bound since $\sum_{j=1}^{\infty} c_j$ is convergent by definition. Defining $C(\delta) = \sum_{j=1} c_j$ and using the above gives a bound on losses due to sampling of $\mathcal{T}(t)$:

$$\lim_{T\to\infty}\sum_{t=1}^{T}(\mu_t^* - \sum_{j=1}^{N(t)}\mu_{jt}P'_{jt})$$

$$\le \frac{P_{Eq}}{K_q}\, r_2(1+\frac{1}{k_2})\, C(\delta)B(\delta).$$

Thus both parts of the theorem have been proved.

__Comments on Application__

To see the use of the theorem in a standard problem area, consider the 2-armed (N-armed) bandit problem. At each point in time a plan can choose one of N random variables $(X_1, X_2, \ldots, X_N)$ where X_j pays one unit with (the unknown) probability P_j. That is, $\mathcal{E}$ can be represented by the set of N-tuples $\{(p_1, p_2, \ldots, p_N) \ni 0 \le p_j \le 1\}$.

To define the (observed) loss function let n_{jt} be the number of choices of X_j to time t and let S_{jt} be the total units of payoff accumulated thereby.

Define

$$\mu_t^* = \max_j\left\{\frac{S_{jt}}{n_{jt}}\right\}$$

and the corresponding loss as

$$\mu_t^* - \sum_{j=1}^{N} P_{jt}\frac{S_{jt}}{n_{jt}}$$

In this case $\mathcal{T}$ can be restricted to the set of N "pure" plans $(\tau_1, \tau_2, \ldots, \tau_N)$ where τ_j is the plan which simply selects X_j at all times. The "mixed" plan defined by the reproductive algorithm then assigns probability P_{jt} to plan τ_j at time t.

Because $\mathcal{T}$ is finite, the theorem guarantees that, if a sequential reproductive plan is used to determine the choice at each time, the cumulative expected losses will with probability 1 not exceed the bound given at the end of the theorem. This result is considerably stronger than the usual results concerning the N-armed bandit problem, where the algorithms only guarantee asymptotic approach to the optimal average rate of payoff.

While direct applications of this theorem are useful and varied, ranging from control through artificial intelligence and economic planning to genetics, its heuristic aspect should not be overlooked. Because the theorem itself has broad application, it suggests that the framework in which it was proved is a useful way of giving precise definition to the concept of adaptation. The theorem is also a useful guideline to more general applications of the reproductive paradigm recast as a sampling procedure; the efficiency of the procedure particularly indicates its usefulness in time-varying (nonstationary) environments. Some of the complex problems in economic planning, ecology and evolutionary theory, where nonlinearity and nonstationarity are the rule rather than the exception, become a good deal more tractable when tackled from this direction.

__Acknowledgements__

This paper presents one phase of research on adaptation carried out at the Logic of Computers Group at The University of Michigan; the group is currently supported by National Institutes of Health (GM-12236) and National Science Foundation (GJ-519).

SIAM J. Comput.
Vol. 2, No. 2, June 1973

GENETIC ALGORITHMS AND THE OPTIMAL ALLOCATION OF TRIALS*

JOHN H. HOLLAND†

Abstract. This study gives a formal setting to the difficult optimization problems characterized by the conjunction of (1) substantial complexity and initial uncertainty, (2) the necessity of acquiring new information rapidly to reduce the uncertainty, and (3) a requirement that the new information be exploited as acquired so that average performance increases at a rate consistent with the rate of acquisition of information. The setting has as its basis a set $\mathscr{A}$ of structures to be searched or tried and a performance function $\mu:\mathscr{A} \to$ real numbers. Within this setting it is determined how to allocate trials to a set of random variables so as to maximize expected performance. This result is then transformed into a criterion against which to measure the performance of a robust and easily implemented set of algorithms called *reproductive plans*. It is shown that reproductive plans can in fact surpass the criterion because of a phenomenon called *intrinsic parallelism*—a single trial (individual $A \in \mathscr{A}$) simultaneously tests and exploits many random variables.

1. Introduction. There is an extensive and difficult class of optimization problems characterized by:

(1) substantial complexity and initial uncertainty;

(2) the necessity of acquiring new information rapidly to reduce the uncertainty;

(3) a requirement that the new information be exploited as acquired so that average performance increases at a rate consistent with the rate of acquisition of information.

These problems derive from a whole range of long-standing questions, such as the following.

How is the productivity of a plant or process to be improved, while it is operating, when many of the interactions between its variables are unknown?

How does one improve performance in successive plays of a complex game (such as Chess or Go or a management game) when the solution of the game is unknown (and probably too complex to implement even if it were known)?

How does evolution produce increasingly fit organisms under environmental conditions which perforce involve a great deal of uncertainty vis-à-vis individual organisms?

How can the performance of an economy be upgraded when its mechanisms are only partially known and relevant data is incomplete?

In each case rapid improvement in performance is highly desirable (or essential), though the combination of complexity and uncertainty makes a direct approach to optimization unfeasible. Often the complexity and uncertainty are great enough that the optimum will be attained, if at all, only after an extensive

* Received by the editors August 3, 1972.

† Department of Computer and Communication Sciences, University of Michigan, Ann Arbor, Michigan 48104. This research was supported in part by the National Science Foundation under Grant GJ-29989X.

period of trial and calculation. Problems of this kind will be referred to here as "problems of adaptation," a usage similar to that of Tsypkin [8], though broader.

Problems of adaptation can be given a more precise formulation along the following lines. Let $\mathscr{A}$ be the set of objects or *structures* (control policies, game strategies, chromosomes, mixes of goods, etc.) to be searched or tried. Generally $\mathscr{A}$ will be so large that it cannot be tried one at a time over any feasible time period. Let $\mu:\mathscr{A} \to [r_0, r_1]$, where $[r_0, r_1]$ is an interval of real numbers, be a "performance measure" (error, payoff, fitness, utility, etc.) which assigns a level of *performance* $\mu(A)$ to each structure $A \in \mathscr{A}$. Conditions (2) and (3) above then reduce to:

(2′) Obtain new information about μ by trying previously untried structures in $\mathscr{A}$ (it being assumed that the outcome of trying $A \in \mathscr{A}$ is the information $\mu(A)$).

(3′) Assure that $\bar{\mu}_t$, the average of the outcomes of the first t trials, increases rapidly whenever the search of $\mathscr{A}$ reveals an A with $\mu(A) > \bar{\mu}_t$.

With no more formalization than this, a dilemma comes into sharp focus. The rate of search is maximized when each successive trial is of a previously untried $A \in \mathscr{A}$. On the other hand, repeated trials of any A' for which $\mu(A') > \bar{\mu}_t$ will increase $\bar{\mu}_t$ more rapidly. In other words, if the rate of search is maximized, the information cannot be exploited, whereas if information obtained is maximally exploited, no new information is acquired.

The basic problem then is to find a resolution of this dilemma. The simplest precise version of the dilemma arises when we restrict our attention to two random variables ξ_1, ξ_2, defined so that the outcome of a trial of ξ_i is a performance $\mu(\xi_i, t)$. The object then is to discover a procedure for distributing some arbitrary number of trials, N, between ξ_1 and ξ_2 so as to maximize the expected payoff over the N trials. If for each ξ_i we know the mean and variance (μ_i, σ_i) of its distribution, the problem has a trivial solution (namely, allocate all trials to the random variable with maximal mean). The dilemma asserts itself, however, if we inject just a bit more uncertainty. Thus we can know the mean-variance pairs but not which variable is described by which pair; i.e., we know pairs (μ, σ) and (μ', σ') but not which pair describes ξ_1. (This is a version of the much studied 2-armed bandit problem, a prototype of important decision problems. See, for example, Bellman [2] and Hellman and Cover [4].) If it could be determined through observation which of ξ_1 and ξ_2 has the higher mean, then from that point on, all trials could be allocated to that random variable. Unfortunately, unless the distributions are nonoverlapping, no finite number of observations will establish *with certainty* which random variable has the higher mean. Here the tradeoff between gathering information and exploiting it appears in its simplest terms. Gathering information requires trials of *both* random variables, with a consequent decrement in average payoff (because the average performance of one of the random variables is less than maximal). On the other hand, premature exploitation of the apparent best random variable, by allocation of most or all trials thereto, runs the risk of a large loss (because there is a nonzero probability that the apparent best is really second best).

A procedure for allocating trials between ξ_1 and ξ_2 will be said to optimally satisfy the conditions (2′) and (3′) if it maximizes the expected performance over N trials. It should be noted that any increase in the uncertainty, such as not knowing some of the means or variances, can only result in a lower expected performance.

Thus a procedure which solves the given problem yields an upper bound on expected performance under increased uncertainty—a criterion against which to measure the performance of various feasible algorithms.

In these terms the objective of this paper is two-fold. First, making the natural extension to an arbitrary number r of random variables determine an upper bound on expected performance under uncertainty. Second, compare the expected performance of the general class of algorithms known as *reproductive plans* (see Holland [5] and later in this paper) to this criterion. It will be shown that, even under conditions of maximum uncertainty, reproductive plans closely follow the criterion. Moreover, reproductive plans can use a single trial to test many random variables simultaneously, a property designated *intrinsic parallelism*. Thus, reproductive plans can exceed the optimum for one-at-a-time testing of random variables. The latter part of the paper will show how this advantage is attained and that it increases in direct proportion to the number of random variables r.

2. Definition of the problem. Several definitions will have to be added to give precise mathematical form to the problem of searching $\mathscr{A}$ under conditions (1), (2′) and (3′). (This is only a partial formalization of problems of adaptation, sufficient for present purposes—a more complete formulation is given in Holland [5].) First of all, let the elements of $\mathscr{A}$ be represented by strings of length l over a set of symbols $\Sigma = \{\sigma_1, \cdots, \sigma_k\}$; i.e., each $A \in \mathscr{A}$ is represented by (or designates) a string of symbols (alleles, weights, etc.) $\sigma_{i_1}\sigma_{i_2} \cdots \sigma_{i_l}$, where $\sigma_{i_h} \in \Sigma$. For simplicity in what follows, $\mathscr{A}$ will simply be taken to *be* the set of strings (rather than the abstract elements represented by the strings). Let $\sigma_1\square\square \cdots \square$ designate the set of all elements of $\mathscr{A}$ beginning with the symbol σ_1. (For example, $\sigma_1\sigma_1\sigma_4, \sigma_1\sigma_2\sigma_1$, and $\sigma_1\sigma_3\sigma_3$ would belong to $\sigma_1\square\square$, but $\sigma_2\sigma_1\sigma_1$ would not.) More generally, let any string ξ of length l over the augmented set $\Sigma \cup \{\square\} = \{\sigma_1, \cdots, \sigma_k, \square\}$ designate a subset of $\mathscr{A}$ as follows: $A \in \mathscr{A}$ belongs to the subset designated by $\xi = \delta_{i_1}\delta_{i_2} \cdots \delta_{i_l}$ if and only if (i) whenever $\delta_{i_j} \in \Sigma$ the string A has the symbol δ_{i_j} at the jth position, and (ii) whenever $\delta_{i_j} = \square$ any symbol from Σ may occur at the jth position in A. (For example, the strings $\sigma_1\sigma_1\sigma_1\sigma_3$ and $\sigma_3\sigma_1\sigma_2\sigma_3$ belong to $\square\sigma_1\square\sigma_3$ but $\sigma_1\sigma_1\sigma_1\sigma_2$ does not.) The set of $(k + 1)^l$ strings defined over $\Sigma \cup \{\square\}$ will be called the set Ξ of *schemata*; they amount to a decomposition of $\mathscr{A}$ into a large number of subsets based on the representation in terms of Σ.

If now there is a probability distribution P over $\mathscr{A}$, say the probability $P(A)$ that $A \in \mathscr{A}$ will be selected for trial, $\mathscr{A}$ can be treated as a sample space and each schema ξ designates an event on $\mathscr{A}$. Accordingly, the performance measure μ becomes a random variable, the elementary event A occurring with probability $P(A)$ and yielding payoff $\mu(A)$. Moreover, the restriction $\mu|\xi$ of μ to a particular subset ξ is also a random variable, $A \in \xi$ being chosen with probability $(P(A))/(\sum_{A \in \xi} P(A))$ and yielding payoff $\mu(A)$. In what follows, ξ will be of interest only in its role of designating the random variable $\mu|\xi$; therefore ξ will be used to designate both an element of Ξ and the corresponding random variable with sample space ξ, distribution $(P(A))/(\sum_{A \in \xi} P(A))$, and values $\mu(A)$. As a random variable, ξ has a well-defined average μ_ξ and variance σ_ξ^2; intuitively, μ_ξ is the payoff expected when an element of ξ is randomly selected (under the distribution P).

Using the decomposition of $\mathscr{A}$ into random variables gives by P and Ξ, it is possible to formalize the earlier discussion concerning optimal allocation of trials.

For the 2-schemata case, let $n_{(2)}$ be the number of trials allocated to the schema with the lowest *observed* payoff rate at the end of N trials. Let $q(n_{(2)})$ be the probability that the schema with the highest observed payoff rate is actually second best. (In detail $q(n_{(2)})$ is actually a function of $n_{(2)}$, N, (μ, σ), and (μ', σ'); hence the necessity of knowing the mean-variance pairs.) If μ_1 is the mean of ξ_1 and μ_2 the mean of ξ_2, then the expected payoff rate for the allocation $(N - n_{(2)}, n_{(2)})$ is

$$\max(\mu_1, \mu_2) - [(N - n_{(2)})q(n_{(2)}) + n_{(2)}(1 - q(n_{(2)}))] \cdot |\mu_1 - \mu_2| \cdot \frac{1}{N}.$$

An optimum value for $n_{(2)}$ can be obtained from this expression by standard techniques. (The development below is somewhat more complicated since, in general, one cannot guarantee a priori that the random variable with the highest observed payoff rate at the *end* of the N trials will have received a predetermined number of trials by that time.)

Though the derivations are much more intricate, the extension from the 2-schemata case to the r-schemata case is conceptually straightforward. It is possible to obtain useful bounds on the payoff rate as a function of the total number of trials, N, together with bounds on the total number of trials which should be allocated to the schema with the highest observed payoff rate. Because the upper and lower bounds so obtained are close to one another (relative to N), the action of the optimal procedure is pretty clearly defined. Using this information it is possible to define a realizable procedure which approaches the optimum as N increases.

If the r-schemata must be tested one at a time, it is clear that one can do no better than the procedure just outlined. If, on the other hand, information about *several* schemata can be obtained from trial of a single individual, $A \in \mathscr{A}$, the rate of improvement could exceed the optimal rate for the one-schema-at-a-time procedure. It should be remarked at once that, for this improvement to take place, the information must not only be obtained but *used* to generate subsequent individuals ($A \in \mathscr{A}$) for trial—each of which will reveal further information about a variety of schemata. The second part of this paper studies a specific set of reproductive plans, the *genetic plans*, which can do just this. It will be shown that genetic plans follow the general course of the optimal procedure when artificially constrained to one-schema-at-a-time searches, but advance much more rapidly when not so constrained.

3. Optimal allocation of trials. For notational convenience in the 2-schemata case, let ξ_1 be the schema with highest mean, ξ_2 the schema with lowest mean. (The observer, of course, does not know this.) Let $\xi_{(1)}(\tau, N)$ be the schema with the highest *observed* payoff rate (average per trial) after an allocation of N trials according to plan τ; let $\xi_{(2)}(\tau, N)$ designate the schema with lowest *observed* rate. Note that for any number of trials n, $0 \leqq n \leqq N$, allocated to $\xi_{(2)}(\tau, N)$, there is a positive probability, $q(N - n, n)$, that $\xi_{(2)}(\tau, N) \neq \xi_2$ (assuming overlapping distributions). Equivalently, $q(N-n, n)$ is the probability that the observed best is actually second best.

THEOREM 1. *Given N trials to be allocated to two random variables ξ_1 and ξ_2, with means $\mu_1 > \mu_2$ and variances σ_1, σ_2 respectively, the minimum expected loss results when the number of trials allocated $\xi_{(2)}$ is*

$$n^* \sim \left(\frac{\sigma_2}{\mu_1 - \mu_2}\right)^2 \ln\left[\left(\frac{\mu_1 - \mu_2}{\sigma_2}\right)^4 \left(\frac{N^2}{8\pi \ln N^2}\right)\right].$$

The corresponding expected loss per trial is

$$l^*(N) \sim \frac{\sigma_2^2}{(\mu_1 - \mu_2)N}\left[2 + \ln\left[\left(\frac{\mu_1 - \mu_2}{\sigma_2}\right)^4 \left(\frac{N^2}{8\pi \ln N^2}\right)\right]\right].$$

Proof. (Given two arbitrary functions, $Y(t)$ and $Z(t)$, of the same variable t, "$Y(t) \sim Z(t)$" will be used to mean $\lim_{t \to \infty} (Y(t)/Z(t)) = 1$ while "$Y(t) \cong Z(t)$" means that under stated conditions the difference $Y(t) - Z(t)$ is negligible).

In determining the expected payoff rate of a plan τ over N trials, two possible sources of loss must be taken into account: (1) The *observed* best $\xi_{(1)}(\tau, N)$ is really second best, whence the $N - n$ trials given $\xi_{(1)}(\tau, N)$ incur an (expected) cumulative loss $|\mu_1 - \mu_2| \cdot (N - n)$; this occurs with probability $q(N - n, n)$. (2) The observed best is in fact the best, whence the n trials given $\xi_{(2)}(\tau, N)$ incur a loss $|\mu_1 - \mu_2| \cdot n$; this occurs with probability $q(N - n, n)$. The expected loss $l(N)$ over N trials is thus

$$|\mu_1 - \mu_2| \cdot [(N - n)q(N - n, n) + n(1 - q(N - n, n))].$$

In order to select an n which minimizes the expected loss, it is necessary first to write $q(N - n, n)$ as an explicit function of n. To derive this function let S_2 be the sum of the outcomes (payoffs) of n trials of ξ_2 and let S_1 be the corresponding sum for the $N - n$ trials of ξ_1. Then $q(N - n, n)$ is just the probability that $S_2/n < S_1/(N - n)$ or, equivalently, the probability that $S_1/(N - n) - S_2/n < 0$. By the central limit theorem, S_2/n approaches a normal distribution with mean μ_2 and variance σ_2^2/n; similarly, $S_1/(N - n)$ has mean μ_1 and variance $\sigma_1^2/(N - n)$. The distribution of $S_1/(N - n) - S_2/n$ is by definition the sum (convolution) of the distributions of $S_1/(N - n)$ and $-(S_2/n)$; by an elementary theorem (on the convolution of normal distributions), this is a normal distribution with mean $\mu_1 - \mu_2$ and variance $\sigma_1^2/(N - n) + \sigma_2^2/n$. Thus the probability $\Pr\{S_1/(N - n) - S_2/n < 0\}$ is the tail $1 - \Phi(x_0)$ of a normal distribution $\Phi(x)$ in standard form so that

$$x = \frac{y - (\mu_1 - \mu_2)}{\sqrt{\sigma_1^2/(N - n) + \sigma_2^2/n}}$$

and $-x_0$ is the value of x when $y = 0$.

The tail of a normal distribution is well approximated by

$$\Phi(-x) = 1 - \Phi(x) \lesssim \frac{1}{\sqrt{2\pi}} \cdot \frac{e^{-x^2/2}}{x}.$$

Thus

$$q(N-n,n) \lesssim \frac{1}{\sqrt{2\pi}} \cdot \frac{e^{-x_0^2/2}}{x_0}$$

$$= \frac{1}{\sqrt{2\pi}} \frac{\sqrt{\sigma_1^2/(N-n) + \sigma_2^2/n}}{\mu_1 - \mu_2} \exp \frac{1}{2}\left[\frac{-(\mu_1-\mu_2)^2}{\sigma_1^2/(N-n)+\sigma_2^2/n}\right]$$

(from which we see that q is a function of the variances and means as well as the total number of trials, N, and the number of trials, n, given ξ_2). Upon noting that q decreases exponentially as a function of n, it becomes clear that, to minimize loss as N increases, the number of trials allocated the observed best, $N - n$, should be increased dramatically relative to n. This observation (which will be verified in detail shortly) enables us to simplify the expression for x_0. Whatever the value of σ_1, there will be an N_0 such that, for any $N > N_0, \sigma_1^2/(N-n) \ll \sigma_2^2/n$, for n close to its optimal value. (In most cases of interest this occurs even for small numbers of trials since, usually, σ_2 is at worst an order of magnitude or two larger than σ_1.) Using this we see that, for n close to its optimal value,

$$x_0 \lesssim \frac{(\mu_1-\mu_2)\sqrt{n}}{\sigma_2}, \qquad N > N_0.$$

We can now proceed to determine what value of n will minimize the loss $l(n)$ by taking the derivative of l with respect to n:

$$\frac{dl}{dn} = |\mu_1 - \mu_2| \cdot \left[-q + (N-n)\frac{dq}{dn} + 1 - q - n\frac{dq}{dn}\right]$$

$$= |\mu_1 - \mu_2| \cdot \left[(1-2q) + (N-2n)\frac{dq}{dn}\right],$$

where

$$\frac{dq}{dn} \lesssim \frac{1}{\sqrt{2\pi}}\left[-\frac{e^{-x_0^2/2}}{x_0^2} - e^{-x_0^2/2}\right]\frac{dx_0}{dn} = -\left[\frac{q}{x_0} + x_0 q\right]\frac{dx_0}{dn}$$

and

$$\frac{dx_0}{dn} \lesssim \frac{\mu_1-\mu_2}{2\sigma_2\sqrt{n}} = \frac{x_0}{2n}.$$

Thus

$$\frac{dl}{dn} \lesssim |\mu_1 - \mu_2| \cdot \left[(1-2q) - (N-2n)\cdot q \cdot \frac{x_0^2+1}{2n}\right].$$

n^*, the optimal value of n, satisfies $dl/dn = 0$, whence we obtain a bound on n^* as follows:

$$0 \lesssim (1-2q) - \left(\frac{N}{2n^*} - 1\right) \cdot q \cdot (x_0^2 + 1)$$

or

$$\frac{N}{2n^*} - 1 \lesssim \frac{1-2q}{q\cdot(x_0^2+1)}.$$

Noting that $1/(x_0^2+1) \lesssim 1/x_0^2$ and that $1-2q$ rapidly approaches 1 because q decreases exponentially with n, we see that $(N-2n^*)/n^* \lesssim 2/(x_0^2 q)$, where the error rapidly approaches zero as N increases. Thus the observation of the preceding paragraph is verified, the ratio of trials of the observed best to trials of second-best growing exponentially.

Finally, to obtain n^* as an explicit function of N, q must be written in terms of n^*:

$$\frac{N-2n^*}{n^*} \lesssim \frac{2\sqrt{2\pi}\sigma_2}{\mu_1-\mu_2}\cdot\frac{1}{\sqrt{n^*}}\cdot \exp\left[\frac{(\mu_1-\mu_2)^2 n^*}{2\sigma_2^2}\right].$$

Introducing $b = (\mu_1-\mu_2)/\sigma_2$ and $N_1 = N - n^*$ for simplification, we obtain

$$N_1 \lesssim \frac{\sqrt{8\pi}}{b}\cdot \exp\left[\frac{b^2 n^* + \ln n^*}{2}\right]$$

or

$$n^* + \frac{\ln n^*}{b^2} \gtrsim \frac{2}{b^2}\cdot \ln\left(\frac{b}{\sqrt{8\pi}}\cdot N_1\right),$$

where the fact that $(N-2n^*) \sim (N-n^*)$ has been used, with the inequality generally holding as soon as N_1 exceeds n^* by a small integer. (To get numerical bounds on σ_2 when it is not explicity known, note that for bounded payoff (all $A \in \mathscr{A}$, $r_0 \leqq \mu(A) \leqq r_1$) the maximum variance occurs when all payoff is concentrated at the two extremes. That is,

$$P(r_0) = P(r_1) = \tfrac{1}{2} \quad \text{and} \quad \sigma_2^2 \leqq \sigma_{\max}^2 = (\tfrac{1}{2}r_1^2 + \tfrac{1}{2}r_0^2) - (\tfrac{1}{2}r_1 + \tfrac{1}{2}r_0)^2 = \left(\frac{r_1 - r_0}{2}\right)^2.)$$

We obtain a recursion for an ever better approximation to n^* as a function of N_1 by rewriting this as

$$n^* \gtrsim b^{-2} \ln\left[\frac{(bN_1)^2}{8\pi n^*}\right].$$

Thus

$$\begin{aligned} n^* &\gtrsim b^{-2} \ln\left[\frac{(bN_1)^2}{8\pi(b^{-2}\ln((bN_1)^2/8\pi n^*))}\right] \\ &\gtrsim b^{-2} \ln\left[\frac{b^4 N_1^2}{8\pi}\cdot\frac{1}{\ln((bN_1)^2/8\pi) - \ln n^*}\right] \\ &\gtrsim b^{-2} \ln\left[\frac{b^4 N_1^2}{8\pi(\ln N_1^2 - \ln(b^2/8\pi))}\right] \\ &\gtrsim b^{-2} \ln\left[\frac{b^4 N_1^2}{8\pi \ln N_1^2}\right], \end{aligned}$$

where, again, the error rapidly approaches zero as N increases. Finally, where it is desirable to have n^* approximated by an explicit function of N, the steps here can be redone in terms of N/n^*, noting that N_1/n^* rapidly approaches N/n^* as N increases. Then

$$n^* \sim b^{-2} \ln \left[\frac{b^4 N^2}{8\pi \ln N^2} \right],$$

where, still, the error rapidly approaches zero as N increases.

The expected loss per trial $l^*(N)$ when n^* trials have been allocated to $\xi_{(2)}(\tau, N)$ is

$$\begin{aligned} l^*(N) &= \frac{1}{N}|\mu_1 - \mu_2| \cdot [(N - n^*)q(N - n, n^*) + n^*(1 - q(N - n^*, n^*))] \\ &= |\mu_1 - \mu_2| \cdot \left[\frac{N - 2n^*}{N} q(N - n^*, n^*) + \frac{n^*}{N} \right] \\ &\gtrsim |\mu_1 - \mu_2| \cdot \left[\frac{2n^*}{N x_0^2} + \frac{n^*}{N} \right] \\ &\gtrsim \frac{|\mu_1 - \mu_2|}{b^2 N} \cdot \left[2 + \ln \left(\frac{b^4 N^2}{8\pi \ln N^2} \right) \right]. \end{aligned}$$

Q.E.D.

The expression for n^* (and hence the one for $l^*(N)$) was obtained on the assumption that the n^* trials were allocated to $\xi_{(2)}(\tau, N)$. However, there is *no* realizable sequential algorithm which can "foresee" in all cases which of the two schemata will be $\xi_{(2)}(\tau, N)$. There will always be observational sequences wherein *each* schema has a positive probability of being $\xi_{(2)}(\tau, N)$ even after τ has allocated $n > n^*$ trials to one. (For example, τ may have allocated exactly n^* trials to each and must decide where to allocate the next trial even though each schema has a positive probability of being $\xi_{(2)}(\tau, N)$.) Thus, no matter what the plan τ, it will in some cases allocate $n > n^*$ trials to a schema ξ (on the assumption that ξ will turn out to be $\xi_{(1)}(\tau, N)$) only to be confronted with the fact that $\xi = \xi_{(2)}(\tau, N)$. For these sequences the loss will perforce exceed the optimum. Hence $l^*(N)$ is not attainable by any realizable sequential algorithm τ—there will always be outcome sequences which lead τ to allocate too many trials to $\xi_{(2)}(\tau, N)$.

There is, however, a realizable plan τ_0 for which the expected loss per trial $l(\tau_0, N)$ quickly approaches $l^*(N)$; i.e.,

$$\lim_{N \to \infty} \frac{l(\tau_0, N)}{l^*(N)} = 1.$$

τ_0 initially allocates n^* trials to each schema (in any order) and then allocates the remaining $N - 2n^*$ trials to the schema with the highest observed payoff rate at the end of the $2n^*$ trials.

COROLLARY 1.1. *Given N trials, τ_0's expected loss, $l(\tau_0, N)$, approaches the optimum $l^*(N)$. That is, $l(\tau_0, N) \sim l^*(N)$.*

Proof. The expected loss per trial $l(\tau_0, N)$ for τ_0 is determined by applying the earlier discussion of sources of loss to the present case:

$$l(\tau_0, N) = \frac{1}{N}\cdot|\mu_1 - \mu_2|\cdot[(N - n^*)q(n^*, n^*) + n^*(1 - q(n^*, n^*))],$$

where q is the same function as before, but here the probability of error is irrevocably determined after only n^* trials have been allocated to *each* schema; i.e.,

$$q(n^*, n^*) \sim \frac{1}{\sqrt{2\pi}}\,\frac{\sqrt{\sigma_1^2/n^* + \sigma_2^2/n^*}}{\mu_1 - \mu_2}\exp\left[\frac{-(\mu_1 - \mu_2)^2}{\sigma_1^2/n^* + \sigma_2^2/n^*}\right].$$

(Note that n^* is *not* being redetermined for τ_0; n^* is the number of trials determined above.) Rewriting $l(\tau_0, N)$ we have

$$l(\tau_0, N) = |\mu_1 - \mu_2|\cdot\left[\frac{N - 2n^*}{N}q(n^*, n^*) + \frac{n^*}{N}\right].$$

Since, asymptotically, q decreases as rapidly as N^{-1}, it is clear that the second term in the brackets will dominate as N grows. Inspecting the earlier expression for $l^*(N)$ we see the same holds there. Thus, since the two second terms are identical,

$$\lim_{N\to\infty}\frac{l(\tau_0, N)}{l^*(N)} = 1. \qquad \text{Q.E.D.}$$

To treat the case of r schemata we need a new determination of the probability that the observed best is not the schema with the highest mean. To proceed to this determination let the r schemata be $\xi_1, \xi_2, \cdots, \xi_r$ and let $\mu_1 > \mu_2 > \cdots > \mu_r$ (again, without the observer knowing that this ordering holds).

THEOREM 2. *Given N trials to be allocated to r random variables $\{\xi_1, \xi_2, \cdots, \xi_r\}$, with means $\mu_1 > \mu_2 > \cdots > \mu_r$ and variances $\sigma_1, \sigma_2, \cdots, \sigma_r$ respectively, the minimum expected loss per trial $l_r^*(N)$ is bounded above and below by $l''_{N,r}$ and $l'_{N,r}$, respectively, where*

$$l'_{N,r} \sim \frac{(r - 1)\sigma_2^2}{(\mu_1 - \mu_2)N}\left[2 + \ln\left[\left(\frac{\mu_1 - \mu_2}{\sigma_2}\right)^4\left(\frac{N^2}{8\pi(r - 1)^2\ln N^2}\right)\right]\right]$$

and

$$l''_{N,r} \sim \frac{(r - 1)\mu_1\sigma_2^2}{(\mu_1 - \mu_2)^2N}\left[2 + \ln\left[\left(\frac{\mu_1 - \mu_2}{\sigma_2}\right)^4\left(\frac{N^2}{8\pi\ln N^2}\right)\right]\right].$$

Proof. We are interested in the probability q_r that the average of the observations of any $\xi_j, j > 1$, exceeds the average for ξ_1; that is, the probability of error

$$q_r(n_1, \cdots, n_r) = \Pr\left\{\left(\frac{S_2}{n_2} > \frac{S_1}{n_1}\right)\text{ or }\left(\frac{S_3}{n_3} > \frac{S_1}{n_1}\right)\text{ or }\cdots\text{ or }\left(\frac{S_r}{n_r} > \frac{S_1}{n_1}\right)\right\}.$$

When a given number of trials $n_0 = \sum_{i=2}^{r} n_i$ has been allocated to $\xi_2, \xi_3, \cdots, \xi_r$ to minimize the probability of error q_r, that error will clearly be largest if $\mu_2 = \mu_3 = \cdots = \mu_r$. (In other words, when $\mu_j \lneqq \mu_2$, an allocation of $n_j < n_2$ trials to ξ_j will yield

$$\Pr\left\{\left(\frac{S_j}{n_j} > \frac{S_1}{n_1}\right)\right\} < \Pr\left\{\left(\frac{S_2}{n_2} > \frac{S_1}{n_1}\right)\right\};$$

hence for a given number of trials, a greater reduction in q_r can be achieved if the means μ_j are not all equal.) Moreover, for those cases where $\mu_2 = \mu_3 = \cdots = \mu_r$, the worst case occurs when the largest of the variances $\sigma_2, \sigma_3, \cdots, \sigma_r$ is in fact the common variance of each of $\xi_2, \xi_3, \cdots, \xi_r$. Given this worst case, $(\mu_2, \sigma_2) = (\mu_3, \sigma_3) = \cdots = (\mu_r, \sigma_r)$, q_r will be minimized for an allocation of n_0 trials to $\xi_2, \cdots, \xi_r$ if (as nearly as possible) equal numbers of trials are given each schema (since each schema contributes equally to the probability of error).

From these observations we can obtain bounds on q_r. As before, let

$$q(n_1, n_2) = \Pr\left\{\frac{S_2}{n_2} > \frac{S_1}{n_1}\right\}.$$

When $\sum_{i=2}^{r} n_i = (r-1)m$ trials are allocated to $\xi_2, \cdots, \xi_r$ so as to minimize the probability of error, we have

$$\Pr\left\{\left(\frac{S_2}{n_2} > \frac{S_1}{N-(r-1)m}\right) \text{ or} \cdots \text{or } \left(\frac{S_r}{n_r} > \frac{S_1}{N-(r-1)m}\right)\right\}$$

$$< \binom{r-1}{1} q(N-(r-1)m, m) - \binom{r-1}{1} 2!\, q(N-(r-1)m, m) + \cdots,$$

where the right-hand side is obtained by noting that the events $\{(S_i/n_i > S_1/(N-(r-1)m)), i = 2, \cdots, r\}$ are independent and, under the best allocation in the worst case, $n_i = m$ for $i = 2, \cdots, r$, so that

$$\Pr\left\{\frac{S_i}{m} > \frac{S_1}{N-(r-1)m}\right\} = q(N-(r-1)m, m).$$

Thus, when $(r-1)m$ trials are allocated to $\xi_2, \cdots, \xi_r$ to minimize the probability of error, we have the following bounds:

$$q(N-(r-1)m, m) < q_r(n_1, \cdots, n_r)$$

$$= \Pr\left\{\left(\frac{S_2}{n_2} > \frac{S_1}{N-(r-1)m}\right) \text{ or} \cdots \text{or } \left(\frac{S_r}{n_r} > \frac{S_1}{N-(r-1)m}\right)\right\}$$

$$< (r-1)q(N-(r-1)m, m).$$

Using the upper and lower bounds on q_r thus obtained, we can proceed to upper and lower bounds, $l^{\mathrm{II}}_{N,r}$ and $l^{\mathrm{I}}_{N,r}$ respectively, on the expected loss $l_N(n_2, \cdots, n_r)$ for N trials:

$$l^{\mathrm{I}}_{N,r}(m) = (\mu_1 - \mu_2)[(N-(r-1)m)q + (r-1)m(1-q)] < l_N(n_2, \cdots, n_r) < l^{\mathrm{II}}_{N,r}(m)$$

$$= \mu_1[(N-(r-1)m)(r-1)q + (r-1)m(1-(r-1)q)],$$

using the earlier discussion of sources of loss and the fact that $(\mu_1 - \mu_2) < (\mu_1 - \mu_j) < \mu_1$ for all j. As before we can determine the optimal value of m for these bounds, m^{**} and m^* respectively, by setting $dl/dm = 0$. Letting $q'' = (r-1)q$ and $q' = q$, we have

$$\frac{dl^{(i)}}{dm} = \mu^{(i)}\left[(N - 2(r-1)m)\frac{dq^{(i)}}{dm} - 2(r-1)q^{(i)} + (r-1)\right] = 0.$$

Or, noting that $1 - 2q^{(i)}$ rapidly approaches 1, we have

$$m^{(i)} \sim \frac{N}{2(r-1)} + \frac{1}{2}\left(\frac{dq^{(i)}}{dm}\right)^{-1}.$$

$q^{(i)}$ decreases with m, so dq/dm is a negative quantity. Since $dq''/dm = (r-1)dq'/dm$, we have $dq''/dm < dq_r/dm < dq'/dm$ and

$$m^{**} \sim \frac{N}{2(r-1)} + \frac{1}{2}\left(\frac{dq''}{dm}\right)^{-1} = \frac{N}{2(r-1)} + \frac{1}{2(r-1)}\left(\frac{dq'}{dm}\right)^{-1}$$

$$> \frac{N}{2(r-1)} + \frac{1}{2}\left(\frac{dq'}{dm}\right)^{-1} \sim m^{*}.$$

That is, $(r-1)m^{**}(N) > n_{\text{opt}} = \sum_{i=2}^{r} n_{i,\text{opt}} > (r-1)m^{*}(N)$. Thus by determining the optimal m for each of the bounds we obtain bounds on n_{opt}, the number of trials which should be allocated to schemata other than ξ_1 in order to minimize expected loss.

m^* is directly obtained from the previous 2-schemata derivation by using $(r-1)m$ in place of n and taking the derivative of q with respect to m instead of n. The result is

$$m^{*} \sim b^{-2}\ln\left(\frac{b^4N_1^2}{8\pi(r-1)^2\ln N_1^2}\right) \sim b^{-2}\ln\left(\frac{b^4N^2}{8\pi(r-1)^2\ln N^2}\right),$$

where $N_1 = N - (r-1)m$ now.

m^{**} is similarly obtained using $(r-1)q$ for q throughout. The result is

$$m^{**} \sim b^{-2}\ln\left(\frac{b^4N^2}{8\pi\ln N^2}\right) \sim m^{*} + \frac{2\ln(r-1)}{b^2}.$$

The corresponding upper and lower bounds on the expected loss per trial are

$$l''_{N,r}(m^{**}) \sim \mu_1\cdot\frac{r-1}{b^2N}\left[2 + \ln\left(\frac{b^4N^2}{8\pi\ln N^2}\right)\right]$$

and

$$l'_{N,r}(m^{*}) \sim (\mu_1 - \mu_2)\cdot\frac{r-1}{b^2N}\left[2 + \ln\left(\frac{b^4N^2}{8\pi(r-1)^2\ln N^2}\right)\right]. \qquad \text{Q.E.D.}$$

4. Allocation of trials by genetic plans. We now have bounds on the best possible performance (in terms of minimizing expected loss) of any plan which tests one random variable (schema) at a time. The objective now is to obtain a measure of the performance of genetic plans in similar circumstances so that a comparison can be made with this criterion. This comparison will reveal two things: (1) Even when the genetic plan is constrained to test one schema at a time, losses decrease at a rate proportional to that decreed by the criterion (though, initially, the plan does not have information about the means and variances required to calculate an optimal allocation of trials). (2) Intrinsic parallelism (tests of many schemata with a single trial) is used to advantage by genetic plans, enabling them to greatly surpass the one-schema-at-a-time criterion. Because both of these points came through convincingly under approximations less severe than "$\sim$", weaker approximations will be used wherever they substantially simplify the derivation.

Specifically, let us consider reproductive plans using genetic operators on a nonincreasing population (i.e., for all t, the average effective payoff rate of the population, $\bar{\mu}'_t$, is 1). Such plans can be diagrammed as in Diagram 1. The genetic operators, $\omega \in \Omega$, of step 7 are either of the form

$$\omega : \mathscr{A} \times \mathscr{A} \rightarrow \mathscr{A} \times \mathscr{A}$$

or else

$$\omega : \mathscr{A} \rightarrow \mathscr{A}.$$

DIAGRAM 1

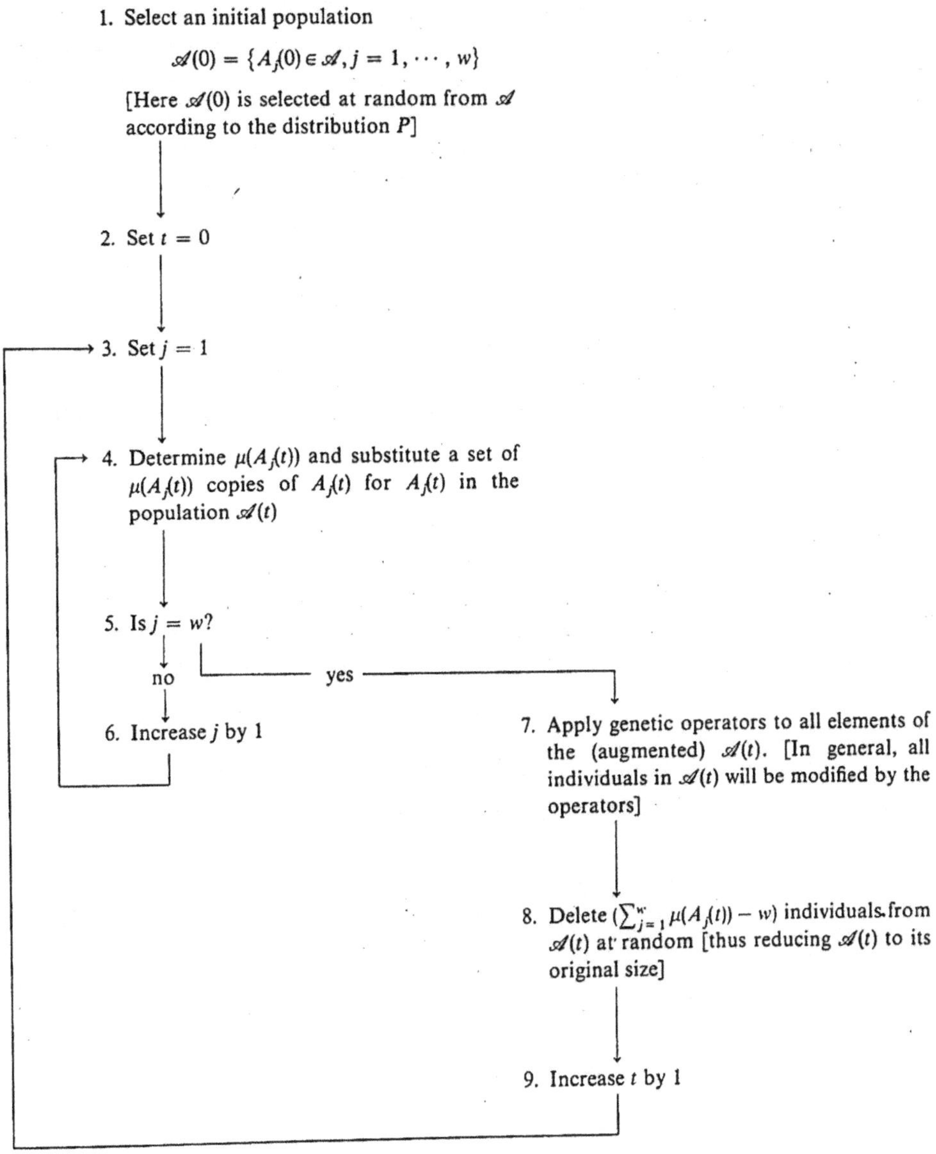
1. Select an initial population
$\mathscr{A}(0) = \{A_j(0) \in \mathscr{A}, j = 1, \cdots, w\}$
[Here $\mathscr{A}(0)$ is selected at random from $\mathscr{A}$ according to the distribution P]
2. Set $t = 0$
3. Set $j = 1$
4. Determine $\mu(A_j(t))$ and substitute a set of $\mu(A_j(t))$ copies of $A_j(t)$ for $A_j(t)$ in the population $\mathscr{A}(t)$
5. Is $j = w$?
no
yes
6. Increase j by 1
7. Apply genetic operators to all elements of the (augmented) $\mathscr{A}(t)$. [In general, all individuals in $\mathscr{A}(t)$ will be modified by the operators]
8. Delete $(\sum_{j=1}^{w} \mu(A_j(t)) - w)$ individuals from $\mathscr{A}(t)$ at random [thus reducing $\mathscr{A}(t)$ to its original size]
9. Increase t by 1

The intended interpretation is that (pairs of) individuals selected from the population $\mathscr{A}(t)$ are transformed by the operator into new (pairs of) individuals. The operators are conservative in the sense that they do not alter the size of the population. Formal definitions of various genetic operators can be found in Holland [6], but for the analysis below, it is necessary to know only that (i) arguments for the operators are chosen at random from $\mathscr{A}(t)$, and (ii) the conditional probability $o_{\xi t}$ that $A \in \xi$, once selected, will be transformed to some $A' \notin \xi$, is generally small and decreases to a value negligibly different from zero as the proportion of ξ in $\mathscr{A}(t)$ approaches 1.

It should be noted that the reproductive plan modifies the distribution P (over $\mathscr{A}$) as the number of trials increases. As a consequence the marginal distributions for the schemata of interest $\{\xi\}$, hence the means $\{\mu_\xi\}$, may change as N grows large. However, the central limit theorem holds for sequences of independent random variables with variable distribution as long as they are uniformly bounded. This condition holds for all schemata when there is an upper bound on performance (l.u.b.$_{A \in \mathscr{A}} \{\mu(A)\} < \infty$), and we shall proceed accordingly.

THEOREM 3. *Given N trials, a reproductive plan with genetic operators can be expected to allocate $N_{\xi_{(1)}}$ trials to the schema (random variable) $\xi_{(1)}$ with the best observed average payoff, where*

$$N_{\xi_{(1)}} \cong (>)\frac{N_{\xi_{(1)}}(0)}{\hat{z}_{(1)}} \exp\left[\frac{\hat{z}_{(1)} n_\rho}{n_\rho(0)}\right],$$

with $\hat{z}_{(1)}$ being the average of the logarithms of the observed payoffs for $\xi_{(1)}$, $N_{\xi_{(1)}}(0)$ being the trials allocated to $\xi_{(1)}$ at the outset, and $n_\rho = N - N_{\xi_{(1)}}$.

Proof. The increase in the number of instances of schema ξ during step 4 of the plan is given by

$$\mathscr{N}'_\xi(t) = \sum_{A \in \xi(t)} \mu(A),$$

where $\xi(t)$ is the set of instances of ξ in the population $\mathscr{A}(t)$. The instances of ξ in $\mathscr{A}(t)$ constitute a sample of ξ under the modified distribution P_t holding at time t. The value of $\mathscr{N}'_\xi(t)$ can be written in terms of $\mathscr{N}_\xi(t)$, the expected number of individuals in $\xi(t)$, by using the average

$$\hat{\mu}_{\xi t} \overset{\text{def}}{=} \frac{1}{\mathscr{N}_\xi(t)} \sum_{A \in \xi(t)} \mu(A)$$

of the observations of ξ at time t. In these terms

$$\mathscr{N}'_\xi(t) = \hat{\mu}_{\xi t} \mathscr{N}_\xi(t).$$

We will concentrate here on (typical) reproductive plans wherein the operators in step 7 are applied to elements of $\mathscr{A}(t)$ independently of their identity (representation). As mentioned earlier, the common characteristic of these operators is such that the conditional probability $o_{\xi t}$ of $A \in \mathscr{A}(t)$ being transformed to $A' \notin \xi(t)$ decreases to a negligibly small value as the size of $\xi(t)$ approaches that of $\mathscr{A}(t)$. Hence at the end of step 7 we can expect

$$\mathscr{N}''_\xi(t) \cong (1 - o_{\xi t})\hat{\mu}_{\xi t} \mathscr{N}_\xi(t),$$

where the factor $(1 - o_{\xi t}) \to 1$ as $\mathcal{N}_\xi(t) \to w$. (This ignores new instances of ξ formed by the operators from other $A \notin \xi(t)$.) Finally, after the deletions of step 8 (which are again uniform over $\mathscr{A}(t)$), we expect

$$\mathcal{N}_\xi(t+1) \cong \frac{1}{\bar{\mu}_t}(1 - o_{\xi t})\hat{\mu}_{\xi t}\mathcal{N}_\xi(t),$$

since the expected value of the increase, $\sum_{j=1}^{w} \mu(A_j(t))$, is just $\bar{\mu}_t = \sum_{A \in \mathscr{A}} \mu(A)P_t(A)$, where P_t is the (modified) distribution over $\mathscr{A}$ at time t. Putting this recursion in explicit form we get

$$\begin{aligned}\mathcal{N}_\xi(t+1) &\cong \left(\prod_{t'=0}^{t} \hat{\mu}'_{\xi t'}\right)\mathcal{N}_\xi(0) = \mathcal{N}_\xi(0) \exp\left[\ln\left(\prod^{t} \hat{\mu}'_{\xi t'}\right)\right]\\ &= \mathcal{N}_\xi(0) \exp\left[\sum^{t} \ln(\hat{\mu}'_{\xi t'})\right]\\ &= \mathcal{N}_\xi(0)\, e^{\hat{z}_t t},\end{aligned}$$

where $\hat{\mu}'_{\xi t} = \hat{\mu}_{\xi t}(1 - o_{\xi t})/\bar{\mu}_t$ and $\hat{z}_t = \sum^{t} \ln(\hat{\mu}'_{\xi t'})/t$. Using the fact that

$$\sum_{t'=1}^{t} f(t') \geqq \int_{t'=0}^{t} f(t')\, dt'$$

for monotone increasing functions, the *total* number of trials of ξ to time t, $N_\xi(t)$, can be approximated by

$$N_\xi(t) \cong \mathcal{N}_\xi(0) + \mathcal{N}_\xi(0) \sum_{t'=1}^{t} e^{\hat{z}_t t'} \geqq \mathcal{N}_\xi(0) + \mathcal{N}_\xi(0) \int_{t'=0}^{t} e^{\hat{z}_t t'}\, dt',$$

assuming that at each time t, $\hat{\mu}'_{\xi t}$ (the rate of increase of $\xi(t)$) is in excess of 1. (It should be noted that this approximation to $N_\xi(t)$ assumes $\hat{z}_1 \cong \hat{z}_2 \cong \cdots \cong \hat{z}_t$. For small t, there may be a considerable error if $\hat{z}_{t'}, t' = 1, \cdots, t$, swings over a wide range, though at worst the error (as a fraction of the true value) will be considerably less than $e^{-\hat{z}_t}$. Moreover, as t increases, $\hat{z}_t$ changes more and more slowly because it is an average, while earlier terms in the sum being approximated are swamped by the larger later terms. Thus the probability of a given error steadily decreases as t increases.)

$$\begin{aligned}N_\xi(t) &\cong \mathcal{N}_\xi(0) + \frac{\mathcal{N}_\xi(0)}{\hat{z}_t}[e^{\hat{z}_t t} - 1]\\ &\cong \mathcal{N}_\xi(0)\left[\frac{e^{\hat{z}_t t}}{\hat{z}_t} + \left(1 - \frac{1}{\hat{z}_t}\right)\right].\end{aligned}$$

Therefore the total number of trials of a schema ξ increases exponentially as a function of time (assuming the performance of ξ is consistently better than the average).

Let $\xi_{(1)}$ be the schema receiving the greatest number of trials over the interval t, and let $\xi_{(1)}(t')$ designate the set of instances of $\xi_{(1)}$ present in the population $\mathscr{A}(t')$ at time t'. Let $n_p(t)$ be the total trials allocated to all other *individuals* $\{\mathscr{A}(t') - \xi_{(1)}(t')\}$ from $t' = 0$ through $t' = t$. Since for all t' the number of individuals in

$\mathscr{A}(t')$ remains constant, the total number of trials $N(t) = N(0) \cdot t$. It follows that

$$\frac{n_\rho(t)}{n_\rho(0)} = \frac{N(t) - N_{\xi_{(1)}}(t)}{N(0) - N_{\xi_{(1)}}(0)} \leq \frac{N(t)}{N(0)} = t.$$

Hence

$$N_{\xi_{(1)}} \cong (>) \mathscr{N}_{\xi_{(1)}}(0)\left[\frac{1}{\hat{z}_{(1)}} \cdot \exp\left(\frac{\hat{z}_{(1)} n_\rho(t)}{n_\rho(0)}\right)\right],$$

where $\hat{z}_{(1)}$ is the observed $\hat{z}_t$ for $\xi_{(1)}$. Or

$$n_\rho \cong (<)(n_\rho(0)/\hat{z}_{(1)}) \ln [\hat{z}_{(1)} N_{\xi_{(1)}}/\mathscr{N}_{\xi_{(1)}}(0)]. \qquad \text{Q.E.D.}$$

The following correspondence allows comparison of this result to the one obtained earlier for optimal allocation.

	"Optimal" [*]	"Reproductive" [ρ]
$N_{\cdot}$, trials allocated to $\xi_{(1)}$	$N_{1*} = N_1$	$N_{1\rho} = N_{\xi_{(1)}}$
$n_{\cdot}$ trials allocated to other schemata	$n_* = (r-1)m^*$	$n_\rho = n_\rho$

Thus we have

$$N_{1*} \sim \frac{(r-1)\sqrt{8\pi}}{b} \exp\left[\frac{1}{2}\left(\frac{b^2 n_*}{r-1}\right) + \frac{1}{2}\ln\left(\frac{n_*}{r-1}\right)\right],$$

where $b = (\mu_1 - \mu_2)/\sigma_2$, vs.

$$N_{1\rho} \cong (>) \frac{N_{1\rho}(0)}{\hat{z}_{(1)}} \exp\left[\frac{\hat{z}_{(1)} n_\rho}{n_\rho(0)}\right],$$

where $\hat{z}_{(1)} = \sum^{t} \ln(\hat{\mu}'_{\xi_{(1)}t'})/t$.

Clearly the two plans behave in roughly the same way, the number of trials allocated to the "best" in each case increasing exponentially as a function of the total number of trials allocated to all other schemata. However, a comparison of expected loss per trial yields much more interesting information. For the reproductive plan the expected loss per trial is bounded above by

$$l''_\rho = \frac{\mu_1}{N}[N_{1\rho} r' q(N_{1\rho}, n'_\rho) + (1 - r' q(N_{1\rho}, n'_\rho)) n_\rho],$$

where r' is the number of schemata which have received n'_ρ (or more) trials under the reproductive plan. It is critical to what follows that $r' \cdot n'_\rho$ need *not* be equal to n_ρ. Each $A \in \mathscr{A}$ is a trial of 2^l distinct schemata. As $\mathscr{A}(t)$ is transformed into $\mathscr{A}(t+1)$ by the reproductive plan, *each* schema ξ having instances in $\mathscr{A}(t)$ can be expected to have $(1 - o_\omega)\mu_\xi/\bar{\mu}_t$ instances in $\mathscr{A}(t+1)$. Thus, over the course of several time steps, the number of schemata r' receiving n'_ρ trials will be much, much greater than the number of trials n_ρ allocated to *individuals* A (where $A \in \mathscr{A}$ but $A \notin \xi_1$) even when n'_ρ approaches or exceeds n_ρ. (As a simple example consider a set of three trials $\{\sigma_1\sigma_1\sigma_1\sigma_2, \sigma_2\sigma_1\sigma_1\sigma_1, \sigma_2\sigma_2\sigma_2\sigma_1\}$. Each of the 6 schemata $\{\sigma_2\square\square\square, \square\sigma_1\square\square, \square\square\sigma_1\square, \square\square\square\sigma_1, \sigma_2\square\square\sigma_1, \square\sigma_1\sigma_1\square\}$ receives 2 trials, so that for $n'_\rho = 2$ we have $r' = 6$ and $n'_\rho \cdot r' = 12$ though n_ρ is clearly 3 (or less). (See below.) This observation, that generally $r'n'_\rho \gg n_\rho$, is an explicit consequence of the reproductive plan's *intrinsic parallelism* (each trial of an individual $A \in \mathscr{A}$ is a useful trial of a great many schemata).

THEOREM 4. *The ratio of the upper bound on the expected loss per trial for a reproductive plan, l''_ρ, to the corresponding lower bound for optimal allocation, l'_*, varies inversely as the number, r', of schemata being tried. Specifically,*

$$\frac{l''_\rho}{l'_*} \to \frac{(\mu_1 - \mu_2)^2 n_\rho(0)}{2\sigma_2^2 \hat{z}_{(1)}} \left(\frac{1}{r' - 1} \right),$$

where the parameters are as defined in the statements of the previous theorems.

Proof. Substituting the earlier expressions for $N_{1\rho}$ and $q(N_{1\rho}, n'_\rho)$ in l''_ρ, and noting that $(1 - r'q(N_{1\rho}, n'_\rho))n_\rho < n_\rho$, gives

$$l''_\rho \lesssim \frac{\mu_1}{N}\left[\frac{r'N_{1\rho}(0)}{\hat{z}_{(1)} b\sqrt{2\pi}} \exp\left[\frac{\hat{z}_{(1)} n_\rho}{n_\rho(0)} - \frac{b^2 n'_\rho + \ln n'_\rho}{2} \right] + n_\rho \right].$$

If $b^2 n'_\rho/2 \geqq \hat{z}_{(1)} n_\rho / n_\rho(0)$, it is clear that the first term decreases as n_ρ increases, but the second term, n_ρ, increases. In other words, if n'_ρ increases at a rate proportional to the rate of increase of n_ρ, the expected loss per trial will soon depend almost entirely on the second term, as was the case for optimal allocation. Thus, for n'_ρ so specified, we can compare losses by taking the ratio of the respective second terms:

$$\frac{l''_\rho}{l'_*} = \frac{n_\rho}{(r - 1)m^*}.$$

(A quick comparison of the first terms of l''_ρ and l'_* also shows that the above condition on n'_ρ is sufficient to assure that the first term of l''_ρ is always less than the first term of l'_*.) This comparison is conservative in the sense that the *upper* bound on the reproductive plan's losses is compared to the *lower* bound on the optimal allocation's losses.

To proceed, let the reproductive plan's loss per trial over N trials be compared to that of an optimal allocation of N trials to the r'-schemata which received n'_ρ or more trials under the reproductive plan. (It should be noted that the above condition on n'_ρ can be made as weak as desired by simply choosing $n_\rho(0)$ large enough.) Substituting the explicit expressions derived earlier for n_ρ and m^* as a function of N gives

$$\frac{l''_\rho}{l'_*} \lesssim \frac{b^2 n_\rho(0) \ln [\hat{z}_{(1)}(N - n_\rho)/N_{1\rho}(0)]}{(r' - 1)(\ln [b^4 N^2/8\pi(r' - 1)^2 \ln N^2])\hat{z}_{(1)}}.$$

Simplifying and deleting terms which do not affect the direction of the inequality, we get

$$\frac{l''_\rho}{l'_*} \lesssim \frac{b^2 n_\rho(0) \ln (\hat{z}_{(1)} N)}{(r' - 1)\hat{z}_{(1)}[2 \ln b^2 N - \ln (8\pi(r' - 1)^2 \ln N^2)]}.$$

Or, as N grows,

$$\frac{l''_p}{l'_*} \to \frac{b^2 n_p(0)}{2\hat{z}_{(1)}}\left(\frac{1}{r'-1}\right). \qquad \text{Q.E.D.}$$

Thus the reproductive plan effectively exploits its intrinsic parallelism—its losses for a given number of trials N, in relation to an optimal (one-schema-at-a-time) allocation, are reduced by the factor r'. We can get some idea of how large this reduction is by looking more closely at the relation between N, n'_p and r'. This relation in turn is more easily approached if we first look more closely at schemata. A schema will be said to be *defined on* the set of positions $\{j_1, \cdots, j_h\}$ at which $\delta_{i_j} \neq \square$. Given Σ with k symbols, there are k^h distinct schemata defined on any given set of $h \leqq l$ positions; moreover, no matter what set of positions is chosen, *every* $A \in \mathscr{A}$ is an instance of one of these k^h schemata. That is, the set of schemata so defined partitions $\mathscr{A}$, and any distinct set of positions gives rise to a different partition of $\mathscr{A}$. (For example, given the alphabet $\Sigma = \{\sigma_1, \sigma_2\}$ and strings of length $l = 4$, the set of schemata defined on position 1 is $\{\sigma_1\square\square\square, \sigma_2\square\square\square\}$. It is clear that every string in $\mathscr{A}$ begins either with the symbol σ_1 or else the symbol σ_2, hence the given set partitions $\mathscr{A}$. Similarly the set defined on position 2, $\{\square\sigma_1\square\square, \square\sigma_2\square\square\}$, partitions $\mathscr{A}$, and the set defined on positions 2 and 4, $\{\square\sigma_1\square\sigma_1, \square\sigma_1\square\sigma_2, \square\sigma_2\square\sigma_1, \square\sigma_2\square\sigma_2\}$, is still a different partition of $\mathscr{A}$, a refinement of the one just previous.) There are $\binom{l}{h}$ distinct ways of choosing h positions $\{1 \leqq j_1 < j_2 < \cdots < j_h \leqq l\}$ along a string of length l, and h can be any number between 1 and l. Thus there are $\sum_{h=1}^{l} \binom{l}{h} = 2^l$ distinct partitions induced on $\mathscr{A}$ by these sets of schemata. It follows that when the reproductive plan generates N trials, they will be simultaneously distributed over each of these partitions. That is, *each* of the 2^l *sets* of schemata (defined on the 2^l distinct choices of positions) receives N trials.

We can get a *rough* estimate of the number, r', of schemata receiving n'_p or more trials by assuming the N trials are distributed uniformly and independently over each partition. Two factors perturb the estimate: (1) Given a uniform initial distribution P, the reproductive plan will make the distribution increasingly non-uniform as n_p increases. However, until N gets fairly large relative to n_p the departure is small enough to make the estimate useful. (2) When a given schema defined on h positions receives n'_p or more trials, then so must every schemata of which it is a subset. (For example, let $\square\sigma_2\square\sigma_1$ receive 2 trials, say $\{\sigma_2\sigma_2\sigma_2\sigma_1, \sigma_1\sigma_2\sigma_1\sigma_1\}$. These are at the same time trials of $\square\sigma_2\square\square$, and also of $\square\square\square\sigma_1$. Hence $\square\sigma_2\square\square$ and $\square\square\square\sigma_1$ also receive at least 2 trials.) Similarly, if a given schema receives less than n'_p trials, then so will every schema of which it is a superset. These are clearly violations of the assumption of independence. Nevertheless, when N is small relative to k^l (so that only a small fraction of schemata have been tried), departures from independence are small enough to allow a useful estimate. Some thought about the number of dependencies relative to the total number of schemata tried, or a small Monte Carlo simulation, are convincing in this respect. Though the estimate is rough, the value of r' obtained for typical values of N, b, $\hat{z}_{(1)}$, etc. is clearly of the right order of magnitude.

The average number of trials per schema for a set of schemata defined on h positions is N/k^h. Under the assumption of uniform, independent trials, the Poisson distribution gives the number of schemata receiving n'_p or more trials:

$$\sum_{n=n'_p}^{\infty} \left(\frac{N}{k^h}\right)^n \left(\frac{1}{n!}\right) e^{-N/k^h}.$$

There are $\binom{l}{h}$ distinct sets of k^h schemata defined on h positions, so that the number r' of schemata in Ξ, $h = 1, \cdots, l$, receiving at least n'_p trials is then

$$\sum_{h=1}^{l} \binom{l}{k} k^h \sum_{n=n'_p}^{\infty} \left(\frac{N}{k^h}\right)^n \left(\frac{1}{n!}\right) e^{-N/k^h}.$$

This is a very large number as long as n'_p is smaller than $N/2$, as it always would be in practice. Even when N is quite small (so that the estimate is good), the number is substantial. For example, if the representations are of length $l = 32$ with two symbols in Σ (so that $\mathscr{A}$ contains $2^{32} \cong 4 \times 10^9$ elements) and if $N = 16$ with $n'_p = 8$, then $r' > 700$ schemata can be expected to receive in excess of n'_p trials. The numbers chosen here are clearly very conservative—if $N = 32$, $r' > 9000$ for l and n'_p as given; any increase in l produces even more dramatic increases in r'.

The advantages implied by this analysis have been observed in a variety of computer tests (Bagley [1], Cavicchio [3], Hollstien [7]).

5. Conclusion. Intrinsic parallelism in the search of schemata offers a tremendous advantage to any optimization procedure which can exploit it. Reproductive plans with genetic operators (genetic algorithms) are the only procedures so far studied which exhibit this phenomenon. They have the additional desirable properties of easy implementation, compact storage and automatic use of the large amounts of relevant information encountered during operation, and robustness (efficient operation under maximal uncertainty). For these reasons it is recommended that genetic algorithms be given serious consideration whenever a problem of natural or artificial adaptation arises.

REFERENCES

[1] J. D. BAGLEY, *The behavior of adaptive systems which employ genetic and correlation algorithms*, Ph.D. dissertation, University of Michigan, Ann Arbor, 1967.

[2] R. BELLMAN, *Adaptive Control Processes*, Princeton University Press, Princeton, N.J., 1961.

[3] D. J. CAVICCHIO, *Adaptive search using simulated evolution*, Ph.D. dissertation, University of Michigan, Ann Arbor, 1970.

[4] M. E. HELLMAN AND J. M. COVER, *Learning with finite memory*, Ann. Math. Statist., 41 (1970), pp. 765–782.

[5] J. H. HOLLAND, *A new kind of turnpike theorem*, Bull. Amer. Math. Soc., 75 (1969), pp. 1311–1317.

[6] ———, *Processing and processors for schemata*, Associative Information Techniques, E. L. Jacks, ed., Elsevier, New York, 1971, pp. 127–146.

[7] R. B. HOLLSTIEN, *Artificial genetic adaptation in computer control systems*, Ph.D. dissertation, University of Michigan, Ann Arbor, 1971.

[8] YA Z. TSYPKIN, *Adaptation and Learning in Automatic Systems*, Academic Press, New York, 1971.

An Introduction to Intrinsic Parallelism in Genetic Code

J. Holland
University of Michigan Dpt. of Computer
and Communication Sciences
2080 Frieze Building
Ann Arbor, Michigan 4810

In this paper I will try to present an overview of genetic algorithms and intrinsic parallelism by looking at the use of a particular genetic algorithm in the optimization of nonlinear functions. (The reader is referred to Holland [1975] for general theory and applications).

Let us begin with an arbitrary function f from a general domain A to the positive reals, $f: A \to R$. In order to give A some structure (to allow us to search for regularities relative to this structure), we will give it a representation via a set of attribute functions $\delta_i: A \to V_i$, $i = 1, \ldots, \ell$. We can think of the δ_i as detectors which pick out attributes or components of each $A \epsilon A$. (For example, in genetics, V_i would be the set of alleles for the ith gene; for the binary expansion of a real number $V_i = \{0,1\}$ for all i; etc.) $A \epsilon A$ has attribute $v \epsilon V_i$ just in case $\delta_i(A) = v$. The representation of A is given by the ℓ-tuple

$$(\delta_1(A), \delta_2(A), \ldots, \delta_\ell(A)).$$

We will assume that A is uniquely characterized in A by its representation. From here onward we will deal only with representations, so A will be taken to be the set of representations.

In these terms f is linear (relative to the representation) when, for all $A \epsilon A$,

$$f(A) = \Sigma_{i=1}^{\ell} \delta_i(A).$$

(For present purposes, broadening the definition to

$$f(A) = \Sigma_{i=1}^{\ell} c_i \delta_i(A)$$

adds no real generality, because this just amounts to a new representation in terms of the detectors $\delta_i' = c_i \delta_i$). When f does have a linear representation, the global optimum is simply obtained because the coordinates can be optimized independently of one another and then summed to yield the global optimum. Unfortunately most functions generated by large systems have no natural linear representation because the systems' components are highly interactive. "Almost all" such nonlinear functions have large numbers of peaks; they are the functions of interest here.

Given a representation, it is instructive to look at sets of elements which have one or more attributes in common. In order to name these subsets of A we will use a special symbol, "$\square$". "$\square$" indicates that we "don't care" what attribute occurs at a given position in the representing ℓ-tuple. Accordingly $(v,\square,\square,\ldots,\square)$ designates the subset of A consisting of all elements possessing attribute $v \varepsilon V_1$ for detector δ_1. Equivalently $(v,\square,\square,\ldots,\square)$ consists of the set of all ℓ-tuples in A beginning with symbol v. Similarly $(\square,v',\square,v'',\square,\square,\ldots,\square)$ designates the subset of A consisting of all elements possessing attribute $v' \varepsilon V_2$ and $v'' \varepsilon V_4$ (and any attribute whatsoever for δ_1, $\delta_3,\delta_5,\delta_6,\delta_7,\ldots,\delta_\ell$). The subsets which can be so named will be called schemata. (They are hyperplanes in the representation space). In general the set of schemata is named by the product set

$$\Xi = \pi_{i=1}^{\ell}\{V_i \cup \{\square\}\}.$$

$A \varepsilon A$ belongs to the schema $\xi = (\Delta_1,\Delta_2,\ldots,\Delta_\ell) \varepsilon \Xi$ if and only if (i) whenever $\Delta_j = v \varepsilon V_j$, $\delta_j(A) = v$ and (ii) whenever $\Delta_j = \square$, $\delta_j(A)$ may have any value whatsoever. That is, whenever $\Delta_j = \square$ we don't care what attribute A has, but whenever Δ_j specifies some attribute $v \varepsilon V_j$, A must have that attribute to belong to the subset $\xi = (\Delta_1,\Delta_2,\ldots,\Delta_\ell)$. It follows that the schema ξ consists exactly of those elements of A having the particular attributes specified by those $\Delta_i \neq \square$ in the ℓ-tuple naming ξ. (The reader is referred to Chapter 4 Holland [1975] for an extended discussion of schemata).

Here the domain A of the function has been given a binary representation to an accuracy of 5 bits. (Only 5 bits are used so that the representations will be short for the purposes of the illustration). Points belonging to the schema $(\square,\square,1,0,\square)$ fall in one of the four shaded strips. That is, the schema $(\square,\square,1,0,\square)$ is the union of the four intervals $1/8 \leqq x < 3/16$, $3/8 \leqq x < 7/16$, $5/8 \leqq x < 11/16$, and $7/8 \leqq x < 15/16$, as can be seen by making all possible substitutions at the three $\square$'s.

We can illustrate the role of schemata in a familiar context by considering a function f of one variable x on the interval $0 \leqq x < 1$.

This is shown in the figure below.

f(.10100) = 1.5
f(.00101) = 1.0
f(.10111) = 0.5
f(.11110) = 1.0
f(.01101) = 1.5

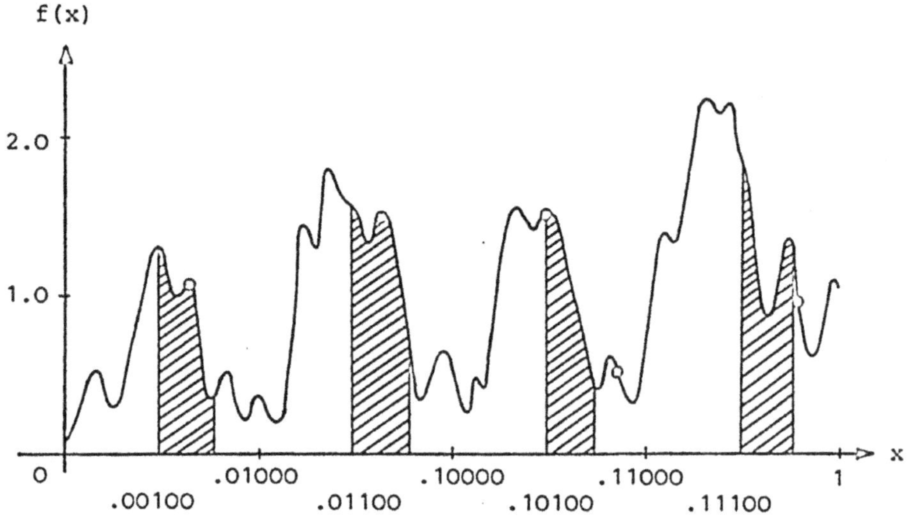

Besinde the plot of f, the values of f for five points drawn at random from A are tabulated. If we think of these points as drawn according to some random distribution P over A, then A becomes a sample space and f becomes a random variable, each AϵA occuring with probability P(A) and yielding value f(A). More importantly under distribution P each schema ξ is a random event and f restricted to ξ, $f|\xi$, is a random variable with a well-defined expectation

$$f_\xi = [\Sigma_{A\epsilon\xi} f(A)\cdot P(A)]/[\Sigma_{A\epsilon\xi} P(A)].$$

Any points drawn from the subset ξ constitute a legitimate sample of the event from which we can form an estimate $\hat{f}_\xi$ of the expectation f_ξ. Thus from the five points tabulated we can form a two point estimate of $f_{(0,\square,\square,\square,\square)}$

$$\hat{f}_{0\square\square\square\square} = (1.0 + 1.5)/2 = 1.25 \ .$$

(Here we abbreviate $(0,\square,\square,\square,\square)$ to $0\,\square\,\square\,\square\,\square$ and assume that the distribution is uniform -- all points are quiprobable. These conventions will hold throughout the example.) Similarly we can form a three point estimate of $f_{1\square\square\square\square}$,

$$\hat{f}_{1\square\square\square\square} = (1.5 + 0.5 + 1.0)/3 = 1.00 ,$$

and a three point estimate of $f_{\square\square10\square}$,

$$\hat{f}_{\square\square10\square} = (1.5 + 1.0 + 1.5)/3 = 1.33 .$$

Because of the higher estimated average for f over $\square\square10\square$, we might consider modifying the distribution so that points are drawn more frequently from this schema than from the other two. More of this later. For now, it is worth noting that each point drawn from A is a legitimate sample point of 2^{ℓ} distinct schemata. (Any schema defined by substituting $\square$'s for one or more of the ℓ attributes in the representation of A is a schema containing A. There are 2^{ℓ} ways of making such substitutions. Even if ℓ is only 20, we obtain information about more than one million schemata each time a point is evaluated!) If this tremendous amount of information can be accumulated and used to direct further sampling, we have a chance of quickly discovering and exploiting regularities in f.

To generalize the foregoing example, consider a "data base" of M elements $\{A_1,\ldots,A_M\}$ drawn from A, with the corresponding observed values of f, $\{f(A_1),\ldots,f(A_M)\}$. We would like to use this information to determine where to draw additional points from A in our search for the global optimum. Let us assume for the moment that the points $\{A_1,\ldots,A_M\}$ are more or less randomly scattered over A. Then one reasonable heuristic suggests that we draw additional points from "regions" in A where above-average values of f have been observed. Of course this heuristic is hopelessly vague without a prior definition of "region". Here we will formalize "region" by schema. Then the heuristic can be given a rigorous formulation as follows.

Let $P_{\xi}(t)$ be the probability of drawing a point (sampling) schema ξ at time t, let $\hat{f}_{\xi}(t)$ be the observed average value of f for the points drawn prior to time t from ξ, and let $\hat{f}(t)$ be the observed average value of f for all points drawn (from A) prior to t. Then, following the heuristic, a change ΔP_{ξ} in the probability of sampling ξ would be determined by the difference $\hat{f}_{\xi}(t) - \hat{f}(t)$. That is, if the average value of f over points drawn from ξ is greater than the average of f over all points drawn, P_{ξ} should be increased (and vice-versa). For example, the recursion

$$P_{\xi}(t + 1) = [\hat{f}_{\xi}(t)/\hat{f}(t)]P_{\xi}(t)$$

produces the suggested changes (while automatically assuring that the probabilities sum to 1 over sets of schemata which partition A). Specifically, this recursion yields

$$\Delta P_\xi \overset{\text{df.}}{=} P_\xi(t+1) - P_\xi(t)$$
$$= [\hat{f}_\xi(t) - \hat{f}(t)]P_\xi(t)/\hat{f}(t)$$

The heuristic so-specified can be supported rigorously along the following lines. Consider a finite set of random variables $\{\xi_1,\ldots,\xi_r\}$, each having an arbitrary distribution with a well-defined (but unknown) mean. At each instant t = 1,2,3,... we are allowed to choose (sample) one of the random variables, say ξ_j, and receive as "payoff" the value $f_j(t)$ observed (the value of f for the particular sample point drawn from the sample space underlying ξ_j). Our objective is to maximize the expected sum of the payoffs received after making a sequence of T choices. (That is, we want to use the information received from sampling the random variables to concentrate the samples on the random variable with the highest mean). This objective is attained if the proportion of trials allocated to ξ_j up to time T, $U_j(T)$, approximates

$$c \cdot \exp[\hat{f}_j(T) - \hat{f}(T)]$$

where c is a constant (based on the means and variances of the random variables). (Theorem 5.3 of Holland [1975] gives a precise statement of this result). The result holds for arbitrary random variables -- that is, arbitrary f's and distributions -- as long as the means and variances are defined.

It is easy to show that the proportions $U_j(T)$ will be produced (with probability 1, to arbitrary accuracy with increasing T) if the probability $P_j(t)$ of sampling ξ_j at time t changes according to the difference equation

$$\Delta P_j \overset{\text{df.}}{=} P_j(t+1) - P_j(t)$$
$$= [\hat{f}_j(t) - \hat{f}(t)]P_j(t)/\hat{f}(t) .$$

(See p. 137 of Holland [1975]). But this is exactly the rate of change of P_j specified by the heuristic, applied now to the random variables $\{\xi_j\}$ as counterparts of the schemata $\{\xi \subset A\}$.

Thus, if we can look upon the schemata $\{\xi\}$ as random variables and sample them with probabilities given by the algorithmic version of the heuristic

$$P_\xi(t + 1) = \hat{f}_\xi(t)/\hat{f}(t)P_\xi(t) ,$$

we will optimize the expected value of the sum of values of f. (This optimization is relative to any plan taking the same number of samples from the set of schemata $\{\xi\}$, regardless of the form of f; it is subject to the usual stochastic qualifications of "occurence with probability 1", etc). This is not to say that the heuristic, as given, is the most efficient way of locating the global optimum. However, it cannot be too far off. Ultimately the expected sum can only be optimized if the point corresponding to the global optimum is frequently sampled. Moreover the exponential concentration of trials (the set of proportions U_j) makes it clear that this time cannot be long delayed (relative to the time it takes to build confidence in estimates of f_ξ). It is also true that for many purposes, optimization of the expected sum is preferable to optimization of the expected maximum. If the search will be a long one (say, of unknown length), it is often important that information be exploited as uncovered, so that the average performance of the algorithm (the average value of points sampled) not remain low for an indefinite period. This amounts to optimization of the expected sum.

(Evolutionary processes and on-line control both illustrate this latter point. An evolving taxon must steadily increase its fitness (average performance) if it is not to have its search ended by extinction. Similarly an on-line control system must at least maintain its current performance while searching for better, if unreasonable costs are to be avoided.)

At this point we know how an algorithm should proceed if it is to optimize the expected value of samples drawn from an arbitrary finite set of schemata (looked upon here as random variables with means and variances defined). We must now show, by construction, that there exist algorithms capable of this behavior. The algorithms we will consider, genetic algorithms, are broadly defined by the following four step iteration:

1) Determine $f(A_i)$ for each element in the data base B(t).

2) Produce a revised data base B'(t) which contains $cf(A_i)$ copies of each element A_i in the given data base B(t).
   (c is chosen to normalize f; e.g. $c = M/\Sigma_i f(A_i)$. The problem of making a "fractional copy", when $cf(A_i)$ is not a whole

number, is handled by constructing an additonal copy with probability equal to the fraction. Thus if $cf(A_i) = 2.4$, a third copy of A_i is made with probability 0.4.)

3) Apply generalized genetic operators to the elements of B'(t) to produce a new data base B(t + 1).
   (The genetic operators are conservative in the sense that B(t + 1) will have the same number of elements as B'(t). In general the elements of B(t + 1) will be different from those in B(t).)

4) Return to step (1).

To determine the effect of this algorithm on schemata let us first consider the effect of the first two steps. Let ξ be a schema with a set of instances $B_\xi(t)$ in B(t). After step (2) there will be

$$c\Sigma_{A_i \varepsilon B_\xi} f(A_i)$$

instances of ξ in B'(t). The average value of f over instances of ξ in B(t) is, by definition,

$$\hat{f}_\xi = \Sigma_{A_i \varepsilon B_\xi} f(A_i)/|B_\xi| ,$$

where $|B_\xi|$ is the number of instances of ξ in B(t). Note that $|B_\xi|/|B| = |B_\xi|/M = P_\xi$, the proportion of the instances of ξ in B(t). Similarly the proportion of instances of ξ in B'(t) is

$$P'(\xi) = c\Sigma_{A_i \varepsilon B_\xi} f(A_i)/c\Sigma_{A_i \varepsilon B} f(A_i)$$

$$= [\hat{f}_\xi \cdot |B_\xi|]/[\hat{f} \cdot |B|]$$

$$= (\hat{f}_\xi/\hat{f})\ (|B_\xi|/|B|) = (\hat{f}_\xi/\hat{f})P_\xi$$

Thus, the change in the proportion of ξ is

$$\Delta P(\xi) = P'(\xi) - P(\xi) = [(\hat{f}_\xi/\hat{f}) - 1]P_\xi$$

$$= [\hat{f}_\xi - \hat{f}]P_\xi/\hat{f} .$$

This is just the action desired. The only problem is that steps (1) & (2) introduce no new instances of ξ. This is where step (3) enters. Genetic operators produce new points while only producing minor perturbations in the action of the first two steps (at least for schemata with two or more instances in the data base). That is, the operators generate new instances of the schemata while assuring the critical rate of increase

$$\Delta P_\xi = (\hat{f}_\xi - \hat{f})P_\xi(1 - \varepsilon_\xi)/\hat{f}$$

where ε_ξ is close to zero for schemata with multiple instances.

The most important of the genetic operators is crossover. Its acts on the data base to produce instances of schemata not previously present in the data base (thus providing new subsets for testing) and it provides new instances of schemata already present (thus increasing confidence in the estimates of the associated f_ξ). Moreover each application of this operator affects a total of $2 \cdot 2^\ell$ schemata in one of these two ways. The operator, a stochastic operator, is defined as follows:

1) Two elements $A = (a_1, a_2, \ldots, a_\ell)$ and $A' = (a'_1, a'_2, \ldots, a'_\ell)$ are selected at random (all elements equilikely) from the revised data base B'(t).

2) A number x is selected with uniform probability from the interval $1 \leq x < \ell$.

3) Two new elements are formed from A and A' by exchanging the attributes of index exceeding x, yielding

   $A'' = (a_1, \ldots, a_x, a'_{x+1}, \ldots, a'_\ell)$

   and

   $A''' = (a'_1, \ldots, a'_x, a_{x+1}, \ldots, a_\ell)$.

By considering the 2^ℓ schemata instanced by A, such as $a_1\square\ldots\square$ and $\square\ldots\square\, a_x a_{x+1}\, \square\ldots\square$, the reader can easily verify the effects of crossover in generating new schemata and new instances. By determining the probability that crossover will destroy an instance of a given schema, without generating a new instance of the same schema, the reader can determine the perturbation ε_ξ on step (2) of the overall algorithm. The details of these considerations, and the effects of other generalized genetic operators, can be found in Chapter 7 of reference [1].

In a data base of M ℓ-tuples randomly chosen from A ($M \ll 2^{\ell/4}$) there will be about $M^2 \cdot 2^{\ell/2}$ schemata with two or more instances. A genetic algorithm will simultaneously process to majority of these schemata in the way dictated by the just-discussed optimization procedure. It does this even though it manipulates elements of the data base rather than directly manipulating schemata. This amounts to a "speed-up" of about $M \cdot 2^{\ell/2}$ over any attempt to treat the schemata one-at-a-time. More specifically, let M_ξ be the number of instances of ξ in the data base, so that $P_\xi = M_\xi/M$ is the proportion on instances of ξ's in the data base. For most ξ for which $M_\xi > 1$, as indicated above, P_ξ will change according to

$$\Delta P_\xi = [\hat{f}_\xi(t) - \hat{f}(t)]P_\xi(t) \ (1 - \varepsilon_\xi)/\hat{f}(t)$$

under a genetic algorithm. (See Lemma 7.2 of Holland [1975]).

It's startling (at least it was to me) that each time a genetic algorithm processes an element of the data base, it implicitly processes nearly $M \cdot 2^{\ell/2}$ schemata in the way specified by the optimization heuristic. This intrinsic parallelism is a great asset in confronting high dimensional, multi-optima problems. Genetic algorithms possess an additional property of importance. After a few iterations of the algorithm, most schemata with two or more instances in the data base B(t), are ranked so that ξ has more instances in B(t) than ξ' if $\hat{f}_\xi < \hat{f}_{\xi'}$. That is, each schema ξ in B(t) is assigned a rank between 1 and M (recalling that M is the size of B) and this rank can be determined for ξ by simply counting the number of instances of ξ in B(t). (See Section 7.5 of Holland [1975]). Thus the genetic algorithm also compactly stores, and provides ready access to, information about a great many schemata --M ℓ-tuples store the relative values of nearly $M \cdot 2^{\ell/2}$ schemata.

Reference

Holland, J.H. 1975. Adaption in Natural and Artifical Systems. University of Michigan Press, Ann Arbor.

A. Lindenmayer and G. Rozenberg (eds.), Automata, Languages, Development

STUDIES OF THE SPONTANEOUS EMERGENCE OF SELF-REPLICATING SYSTEMS USING CELLULAR AUTOMATA AND FORMAL GRAMMARS

John H. Holland
Logic of Computers Group
Computer & Communication Sciences Department
The University of Michigan
Ann Arbor, Michigan 48104

The thesis that life can originate spontaneously has a pedigree going back to Greek times and beyond. Aside from its great intrinsic interest, the thesis has a pivotal role in the study of evolution. For evolution to begin, self-replicating systems must emerge spontaneously from a physical medium exhibiting much lower levels of organization (unless the medium is "seeded" from elsewhere). Many current challenges to the neo-Darwinian interpretation of evolution are based on calculations of the improbability of this emergence. (See, for example, Mathematical Challenges to the Neo-Darwinian Interpretation of Evolution, ed. Moorhead, P. S., and Kaplan, M. M., Wistar Institute Press, Philadelphia, 1967). Before self-replicating systems emerge there exists only the "primordial soup"; after emergence the way is open for prodigious increases in organizational variety and complexity. It seems clear that the heritable adaptations which follow, and thereby the whole of evolution, will be strongly influenced by early steps in this process.

Under the stimulus of a succession of experimental refutations, the thesis of spontaneous emergence has become progressively more sophisticated until, recently, the thesis has acquired a strong experimental base at the molecular level. However, it still lacks a deductive framework and, in particular, it lacks rigorous proofs establishing the existence and properties of various idealized "spontaneous emergence" processes. Lacking such proofs the thesis is in the position of a physics of gravitation without its idealized "free fall" models -- there is no starting point for more elaborate models which approximate observable processes and yield deductions about them. The object of the present study is to provide some of these existence proofs.

The approach is via a two step procedure. First a range of model "universes" having abstract counterparts of basic kinetic and biological operators (such as diffusion and enzymes) is defined. Then, for arbitrary initial conditions in these universes, an analysis is made of the probabilities of emergence of self-replicating systems.

The model universes are unabashedly artificial, but they have been selected with a careful eye to established properties of natural systems. Progressive elaborations of the basic model should yield fair approximations, and simulations, of natural systems without altering the basic points proved for the original model. In broad outline, the model universes have the following features.

1) Structures are defined by combinations of elements drawn from a basic "alphabet" (cf. nucleotides or amino acids).

2) Elements are treated as persistent and readily available.

3) All structures are subjected to the same "kinetic" operators (cf. diffusion and activation) which incessantly cause the separation and combination of elements.

This research was supported in part by National Science Foundation Grant No. DCR71-01997.

4) Some definable structures correspond to biological operators (such as enzymes) which can affect local combination rates.

5) "Starting" conditions are randomized homogeneous mixtures of elements.

Mathematical specification of these conditions makes it possible to demonstrate that, in the corresponding universes, self-replicating systems emerge in a time which is minute compared to the time required under unbiased random generation.

1. INFORMAL DESCRIPTION OF α-UNIVERSE PHYSICS

The basic "physics" of each model is specified by a cellular automaton (see Burks 1970). (The reader need have no prior knowledge of cellular automata since only elementary properties will be required here, and these will be developed as needed). In effect, the underlying physical space is treated as a regular tesselation of cells (the "geometry"), so that each cell has its neighbors arrayed in the same way -- they all fit the same "neighborhood template". In this tesselation, each cell is in one of a finite number of states at any given instant t, the set of possible states being the same for all cells. The cell's state at time t+1 is completely determined (up to random, noise-like variations) by its state and that of a specified set of neighboring cells at time t. Moreover, the algorithm for determining the next state, the transition function, is the same for all cells. The result is a uniform (discrete) geometry and a set of laws which hold at each point in that geometry (cf. the basic laws of physics).

For present purposes the state of each cell will indicate whether the cell is occupied by some one of a set X of basic elements and, if so, how that element is linked to elements in various neighboring cells. The cells of the space should not be confused with biological cells -- the cells of the space are merely locations which can be occupied by a basic element. Similarly, the basic elements should not be confused with atoms -- they may be amino acids, enzymes or whatever other biomolecules serve best as primitives for a given study. The linkage of elements (across cells) serves to create complex compounds from the elements. Thus, the state of a cell is given by a pair of letters (x,y), where $x \varepsilon X$ indicates which basic element (or no element at all) is present, and $y \varepsilon Y$ indicates the bonds that element has formed with neighboring elements.

As in physics, changes of state are determined by the actions and interactions of a set of operators. That is, the transition function is given implicitly. Two broad classes of operators can be distinguished in these investigations: primitive (or a priori) operators and emergent operators. The primitive operators are those which are effective from the outset, such as diffusion operators (providing random walks of the basic elements from cell to cell) and bond modification operators. The emergent operators influence the course of events only after certain compounds of the basic elements have formed, such as compounds which are catalytic with respect to the formation of other compounds. The rules for applying the operators, both primitive and emergent, are completely determined by the configurations of elements in the cells (the cell states), though the operators may have a random component. Thus, as mentioned earlier, the transition function depends only upon the cell states up to a random variation.

The particular models formalized here, the α-universes, constitute only a small fragment of the models just described. The α-universes are a "proof-of-principle" set, intended to demonstrate the interest and relevance of the full class. They employ only the most austere counterparts of important chemical and biochemical mechanisms. For example, a single two-level bond stands in place of the rich variety of bonds present in natural systems. In making such restrictions, we lose much of the texture of the intricate set of interactions mediated by the full panoply of biomolecular mechanisms, but there are two compensations. We see how little is required for the emergence of hereditary adaptive systems, and we see something of the skeleton of the process.

The set X of basic elements for an α-universe is partitioned into three subsets A, N, and {-}. The elements of A correspond to "functional" primitives (cf. amino acids), while the elements of N correspond to "codons" (cf. nucleotide triples). (As developed below the elements of A actually correspond more closely to selected polypeptides and the elements of N to codon sequences). The null element "-" is used to indicate an empty cell.

$$X = A \cup N \cup \{-\}$$

$$A = \{0,1,:\}$$

$$N = \{N_0, N_1\}$$

Sequences over {0,1} designate structures, specifically emergent operators with their arguments; the : is a punctuation used to separate the sequences into meaningful segments, such as the part which designates the type of operator, the parts designating the arguments, etc. (See sec. 2 for particular examples).

Combinations of letters from N encode (or name) the letters in A according to the following table.

code	letter coded
N_0N_0	0
N_0N_1	1
N_1N_0	:

Some emergent operators decode N-strings to produce A-strings. There are interesting possibilities, even at this level, for studying the emergence of a code, but for the present "bare bones" model the coding will be taken as fixed at the outset.

The α-universes will employ only two primitive operators: the EXCHANGE operator (producing diffusion-like movements of the basic elements) and the BOND-MODIFICATION operator (causing random changes in the level of bonds between elements). As for emergent operators, the investigation will also center on two: a COPY operator, and a DECODE operator. The COPY operator attaches itself to a linked string of elements in the N-alphabet (a string which satisfies the COPY operator's arguments) and produces a copy of that string. (The action is similar to a polynucleotide polymerase). The DECODE operator takes a linked string in the N-alphabet and produces a string in the A-alphabet according to the coding table above. (This is a counterpart of the ribosomal transcription of a polynucleotide to a polypeptide).

Because the operators deal with strings, the "proof-of-principle" derivations can be carried out in a one-dimensional cellular space. This simplifies some of the derivations substantially, and also provides contact with "grammatical" approaches to development (see Lindenmayer, 1974).

(Extensions to two- and three-dimensional spaces, and to higher-dimensional "phase" spaces, along with other operators, will be discussed later).

The state S(t) of a one-dimensional α-universe at any time t is given by a string of pairs of symbols. Each pair in the string designates the state of the cell at the corresponding position (reading from left to right) in the cellular array. In particular, the first component of the pair specifies what element is present and the second component indicates the form of the bond between that element and the elements in adjacent cells. The bond we will consider has two levels, <u>weak</u> and <u>strong</u>, and is symmetrical (i.e. if an element is linked to the element at its right, then that element on the right is linked by the same bond to the given element). For convenience the form of the bond will be indicated in only one of the cells involved. Specifically, each cell will indicate what form of bond (strong or weak) exists between the element it contains (if any) and that in the cell on its right. (By convention the bond between an element and an empty cell is a weak bond.) Using Y = {s,w} to represent bond type (strong and weak, in the order given) we have the state of each cell given by the pair (x,y), as indicated earlier, with $x \in X = A \cup N \cup \{-\}$ and $y \in Y = \{s,w\}$.

The α-universes are conservative in the sense that elements are never created or destroyed. They are only moved about and rearranged by the operators. Thus the number (or global proportion) of each type of element is a parameter of the universe. The basic structures of the universe are linked substrings over $A \cup N$, separated by empty cells from other elements or substrings in the universe. These will be called free substrings. Some free substrings, as indicated earlier, will be interpretable as emergent operators. Viewed as a grammar the α-universe has free substrings as its basic interpretable sentences, and it generates these sentences by rearranging words (elements) rather than by creating them at need.

In the early stages of the α-universe's history, when emergent operators are rare or non-existent, rearrangement is a stochastic process mediated by the primitive operators, EXCHANGE and BOND-MODIFICATION. Their interaction produces mixing and changes of linkage which are the counterparts of diffusion and random activation (by collision or energetic photons) in physical systems. These interactions are not element specific (or context sensitive); all elements have an equal chance of being affected.

Eventually, the rearrangements effected by the primitive operators produce interpretable free substrings. The interpretable substrings, i.e. the emergent operators, produce (or encourage the production of) substrings which would be quite rare under the action of the primitive operators alone. This occurs because the emergent operators are generally very context sensitive. As such they are the counterparts of catalysts, enzymes, and antibodies in physical systems.

When the α-universe is viewed as a grammar, we see that it is a stochastic grammar because of the stochastic operators (production rules); thus, a changing probability distribution over the corpus, rather than the corpus alone, is central to a description of its productions. Moreover, it is a "self-descriptive" grammar, since some of its production rules can be described (and employed) by the grammar itself. These rules (operators), when generated, of course affect the probability distribution over the corpus.

The α-universes have three properties (Ω-properties) we would expect to find in most interesting models of evolving universes:

Ω.1) The stability of a structure (particularly its resistence to being broken down by the primitive operators) is a primary determinant of its density (probability of occurrence) in any given region. (For example, consider two structures A_1 and A_2 which have probabilities $p_1 = p_2$ of formation, but probabilities $q_1 = 0.1\, q_2$ of being broken down. In a space operating strictly according to these probabilities, the density of A_1 will be 10 times that of A_2).

Ω.2) The density of a structure is a primary determinant of the frequency with which that structure is operated upon (serves as an operand). (That is, structures, including emergent operators, are being continually "shuffled" by diffusion operators. Thus, of two structures satisfying a given argument, the one with high density is the more likely to serve as operand.)

Ω.3) There are important operators which, when applied to an operand, produce results which inherit characteristics of the operand (say, contiguous sequences of elements). (In the α-universes, the EXCHANGE operator, among others, has this property).

These properties, taken together, imply that any configuration of elements which is positively correlated with above-average stability will appear more often than other possible configurations in observed structures (free substrings).

To state this more carefully: Let a particular configuration of elements, not necessarily adjacent, within a structure be called a schema (cf. the active site of an enzyme, the combining region of an antibody, etc.), and let a schema positively correlated with above-average stability be called a +-schema. Consider, now, the set of all structures $\mathcal{F}_\xi$ containing a given +-schema ξ. The set $\mathcal{F}_\xi$ will occur at an above-average density. That is, there will be progressive departures from uniform random sampling of schemata (and of structures) if any +-schema exist.

There is an important element of intrinsic parallelism here in that each structure amounts to a test of many configurations (schemata). (In the α-universes, a free substring involving ℓ elements is a test of 2^ℓ distinct configurations). On the average, each configuration is sampled (occurs in various contexts) at a rate appropriate to its contribution to stability. In effect, though structures (free substrings) are being tested, it is substructures (schemata) which are having their densities adjusted. This is a tremendous "speed-up" relative to uniform random trials. (Intrinsic parallelism is discussed at length in Adaptation in Natural and Artificial Systems. See Holland, 1975).

The α-universes are intended, above all, to demonstrate that self-replicating systems can emerge from unorganized initial states. Even though α-universes possess the Ω-properties in an impoverished, near-minimal form, we will see that self-replicating systems emerge in a time which is very short relative to the time required to produce the same system by a sequence of independent random changes. This is primarily a consequence of the intrinsic parallelism in "testing for stability". It is also true that, of two structures equally likely to give rise to a more stable variant, it is the more stable (more common) structure which is likely to give rise to the variant which actually occurs. This leads to "lines of succession" and, because the universe is conservative, a transfer of elements from less stable structures to more stable structures. Thus, even at a pre-biotic stage, we observe varying fitness (differential survival of characteristics influencing stability) and niches (competition for limited resources -- the elements).

2. DEFINITION OF THE α-UNIVERSE OPERATORS

The state S(t+1) of the α-universe at time t+1 is determined from the state S(t) at time t in three phases. During phase I, the BOND-MODIFICATION operator is applied to the string. Each bond in the string S(t) is tested; strong bonds are changed to weak bonds with probability r, and weak bonds are changed to strong bonds with probability λr. That is, at each strong bond a trial is made of a random variable, the strong bond reverting to a weak bond with probability r, and remaining strong otherwise; weak bonds are similarly tried but with probability λr. This constitutes the definition of the BOND MODIFICATION operator.

During phase II, the EXCHANGE operator is applied to the string. Each weak bond has a probability m_1 of serving as the center of the exchange operator. (The trials determining which bonds are to serve as centers are carried out from left to right and the EXCHANGE operator is applied at each center in that order). At each center the outer limits of the EXCHANGE must be determined. Each weak bond to the right of the center, in succession, has a probability m_2 of being chosen as the outer limit. That is, a random trial is made at the first weak bond to the right of the center. With probability m_2 the outcome of that trial will indicate that that bond is the outer limit of the exchange; otherwise the process is repeated at the next weak bond to the right. Once the outer limit on the right is determined, the outer limit on the left is determined by counting the same number of weak bonds to the left of center. The EXCHANGE is then effected by interchanging the segment to the left of the center with the one to the right (the left to right ordering of each segment being preserved).

Before EXCHANGE:

```
                 ,outer limit (2 weak bonds to left of center)
                 !           center (chosen with probability m1)
                 !           !                  outer limit (2 weak bonds to
                 !           !                  !          right of center)
...(:,s)(0,w)'(-,w)(:,w)'(0,s)(1,w)(-,w)'(-,w)(0,s)...
```

After EXCHANGE:

```
...(:,s)(0,w) (0,s)(1,w)(-,w) (-,w)(:,w) (-,w)(0,s)...
```

During *phase III*, the emergent operators (if any) are executed. The search for emergent operators amounts to parsing the string which represents the state of the α-universe; certain substrings are interpretable as operators which affect other parts of the α-universe. For present purposes a substring will be interpretable if the X-components of a sequence of pairs in the α-universe satisfy the following simple format:

$$- \delta_1\delta_2 \ldots \delta_u : \delta_{u+1}\delta_{u+2} \ldots \delta_{u+v} -$$

where $\delta_j \varepsilon \{0,1\}$ $j=1,\ldots,u+v$. Note particularly that the substring must have an empty cell at each end and it must consist only of letters from the A-alphabet. The value of the y component of each pair does not enter into the string's interpretability, but it does affect its persistence (stability), as we shall see.

Once a substring satisfying the foregoing format is located, the parts $\delta_1 \ldots \delta_u$ and $\delta_{u+1} \ldots \delta_{u+v}$ are interpreted, respectively, as a designation of the operator type, and the argument (which specifies the operand to be manipulated by the operator). It may be that some sequences of 1s and 0s do not designate operators -- i.e. they act as null operators -- but this is a matter of definition within the particular study. It may also be that binary, or higher order, operators are desirable; the required additional argument(s) (operand designation(s)) can be provided by using additional colons followed by sequences of 1s and 0s. Thus a binary operator would have the format

$$- \delta_1\delta_2 \ldots \delta_u : \delta_{u+1}\delta_{u+2} \ldots \delta_{u+v} : \delta_{u+v+1}\delta_{u+v+2} \ldots \delta_{u+v+w} - .$$

The operand designation typically specifies a substring (or class of substrings, say those having a given prefix) which the operator will act upon. As indicated previously, we will concentrate on conservative operators -- operators which neither create nor destroy elements, but instead recombine them in various ways. As a consequence, the basic action of each operator will be that of linking selected elements (or substrings) after bring them into adjacency (cf. the action of an enzyme).

In the discussion which follows a substring will be called free if it has an empty cell at each end (and no empty cells internally). The operators of interest only manipulate free substrings and elements.

As a first example consider the execution of the COPY operator. The COPY operator's net action is to attach itself to a free N-alphabet substring, the operand, which has a prefix specified by the operator's argument, and then the operator produces a copy of the operand, freeing both the operand and the copy when finished. In "flow-diagram" form, this can be specified as involving the following major steps:

1) A free : is attached by a strong bond to the right of the operator substring.

- [op. type] : [arg.] - ⇒ - [op. type] : [arg.] : -

(The nearest free : to the operator substring has probability m_2 of being chosen for this purpose; if it is not chosen then the next nearest : has probability m_2 of being chosen for this purpose.

The procedure is the same as that used to determine the outer limit of the EXCHANGE operator. Once a free : has been chosen, say on the right of the operator substring, all the elements in cells between the operator and the chosen : are shifted right one cell and the : is shifted to the vacated cell next to the operator.

$$\ldots x_{-1}[\text{op.}]x_1x_2x_3\ldots x_\ell : x_{\ell+2}$$

$$\Rightarrow \ldots x_{-1}[\text{op.}] : x_1x_2\ldots x_\ell x_{\ell+2} \ldots$$

This procedure is necessitated by the conservative nature of the operator. It is the counterpart of the action of thermal velocities in assuring quick interactions between enzyme and substrate).

2) A free substring, with a prefix specified by the argument of the operator substring, is attached by a strong bond at the right of the augmented operator substring (resulting from step 1)).

(The argument is a string of 1s and 0s, from the A-alphabet, while the operand is intended to be in the N-alphabet. Thus an "anticode" is needed. For present purposes this is given by the table

anticode	N-letter specified
0	N_0
1	N_1

Thus the argument of the COPY operator by convention psecifies a prefix in the N-alphabet. An operand having this prefix (if one is available) is chosen by the same procedure as was used to select the : in step 1), using the same probability m_2, and shifting symbols in the same way once an operand is located. The object here is to provide the operator with an opportunity to contact different operands satisfying its argument -- as would be provided by thermal kinetics.)

3) A free : is attached by a strong bond at the right of the augmented substring.

(Essentially a repeat of step 1)).

4) Free elements, in the N-alphabet, are attached with strong bonds one-by-one, in the order specified by the operand, until a non-N alphabet letter is encountered in the operand, or until the end of the operand is encountered.

(The : added in step 2) is interchanged with the N-letter in the cell on its right when that letter has been "copied", i.e. when a duplicate of the letter has been added at the extreme right. If an appropriate free symbol is not available, the operation "holds" until one is available. Thus, this : serves as a "position marker" indicating what N-letter should be added next. The : added in step 3) marks the end of the operand, separating the operand from the copy of it produced by this step.)

5) The completed copy, the operand, and each of the two attached :s, are freed by the "insertion" of empty cells at appropriate points.

- [op. type] : [arg.] [operand] :: [copy] -

⇒ - [op. type] : [arg.] - [operand] -:-:- [copy] -

(The space - is selected with probability m_2 and symbols are shifted as in the selection of the : in step 1). However, the space is drawn only from contiguous sequences of 2 or more spaces, so that the removal of the space from its initial position will not inadvertently cause two substrings to be joined into one.)

The second emergent operator of interest here is the DECODE operator. The DECODE operator attaches itself to a free N-alphabet substring which has a prefix specified by the argument, and then it produces an A-string by decoding the operand. The detailed procedure is the same as that for the COPY operator (acquisition of :s and operand, step-by-step interpretation of the operand using one : as an advancing position marker, freeing the result) except that in step 4) the result is an A-string produced by decoding successive pairs of the operand (N-string) to determine, in order, what A-elements should be added to the right of the complex. (The "position-marking" : is advanced two cells at a time. Any single N-letter at the end of the operand or preceding an A-letter in a pair, is left uninterpreted and step 4) is terminated.)

To keep the proof-of-principle universe as simple as possible, the single A-letter 0 will designate the COPY operator and the letter 1 will designate the DECODE operator. Hence the free string with X-components

$$- 0 : u_1u_2 \ldots u_\ell - \qquad \text{where } u_j \varepsilon \{0,1\}$$

is a copy operator with argument $u_1u_2 \ldots u_\ell$, and

$$- 1 : u_1u_2 \ldots u_\ell - \qquad u_j \varepsilon \{0,1\}$$

is a DECODE operator with argument $u_1u_2 \ldots u_\ell$.

3. PARAMETERS AND THEOREMS

The definitions of the operators, together with the rules for applying them, completely determine the transition function for an α-universe. Since there are no inputs to an α-universe, specification of an initial state would completely determine subsequent history, were it not for the stochastic nature of the operators. Even so, it is possible to prove strong statements about the evolution of α-universes, statements quite analogous to those of statistical mechanics in standard physics.

For any particular example of an α-universe there will be a set of parameters, again analogous to those of statistical mechanics, which are related by the theorems and which determine details of the evolution. Herewith is a list of parameters, and their analogies, for the specific example developed thus far.

Parameter	Informal Interpretation
ρ	"density of matter"

[$(1-\rho) \overset{df.}{=}$ proportion of null element, -; $\rho(e) \overset{df.}{=}$ proportion of element e; since elements are neither created nor destroyed these proportions are fixed.]

m_1	"mean velocity"
m_2	"mean free path"
r	"stability" of strong bond
λ	"equilibrium ratio" between weak and strong bonds

The following are typical theorems involving these parameters.

Lemma 1. The expected density of structures (free substrings) of length ℓ under a uniform random distribution of elements over cells in the α-universes is

$$D(\ell) > \left[\frac{1}{(1-\rho)^2\rho^\ell} + \frac{1}{(1-\rho)}\right]^{-1} ,$$

with an error always less than $D(2\ell)$,

$$\cong (1-\rho)^2\rho^\ell , \qquad \text{for } (1-\rho)\rho^\ell << 1 .$$

Proof:

(This particular proof could be carried out with less mathematical apparatus, but the procedure is useful for more complicated questions about distributions in cellular spaces).

A structure of length ℓ occurs in a sequence of $\ell+2$ cells if the first and last cells contain null elements (-) and the other ℓ cells contain elements of $A \cup N$. Under a uniform random distribution a null element occurs in any given cell with probability $(1-\rho)$ and an element of $A \cup N$ occurs with probability ρ.. In order to make the occurrence of structures mutually exclusive events (in a given set of cells) we will adopt the Feller (1950) convention that two structures which share a common null element will be counted as one. (Thus, given structures ξ and ξ' of length ℓ, the particular sequence - ξ - ξ' - will be counted as one structure, not two. All other sequences are counted properly; we will see how to compensate for the "aberrant" counts later in the proof.)

Let p_n be the probability that cell n contains a null element which terminates a structure of length ℓ. Then, in a sequence of $\ell+2$ cells containing null elements at the ends and elements of $A \cup N$ elsewhere, we have, because of Feller's convention either the first or else the last cell as the termination of a structure. Because these two events are mutally exclusive, their probabilities add. Using this with the fact that the overall sequence occurs with probability $(1-\rho)^2\rho^\ell$, we get the follwoing recursion for P_n,

$$(1-\rho)^2\rho^\ell = p_n + (1-\rho)\rho^\ell\, p_{n-\ell-1} .$$

Using a fundamental theorem on recurrent events (see Feller 1950) we have as the generating function of the recurrence times of structures of length ℓ

$$G(Z) = 1 - [\Sigma_{n=0}^{\infty} p_n Z^n]^{-1} .$$

Multiplying both sides of the recursion by Z^n and summing both sides over $n \geq \ell$ we obtain the equation

$$\frac{(1-\rho)^2\rho^\ell Z^{\ell+2}}{1-Z} = [(\Sigma_{n=0}^{\infty} p_n Z^n) - 1]\, \frac{1-[(1-\rho)\rho^\ell Z^{\ell+1}]^2}{1-(1-\rho)\rho^\ell Z^{\ell+1}}$$

Solving this equation for $\Sigma_{n=0}^{\infty} p_n Z^n$, substituting to obtain $G(Z)$, and using the fact that the expected value of the corresponding distribution is the value at $Z=1$ of the derivative dG/dZ, we get

$$D^{-1}(\ell) = \frac{dG}{dZ}\Big|_1 = \frac{1-[(1-\rho)\rho^\ell Z^{\ell+1}]^2}{[1-(1-\rho)\rho^\ell Z^{\ell+1}](1-\rho)^2\rho^\ell Z^{\ell+2}}\Big|_1$$

$$= \frac{1-[(1-\rho)\rho^\ell]^2}{[1-(1-\rho)\rho^\ell](1-\rho)^2\rho^\ell} = \frac{1}{(1-\rho)^2\rho^\ell} + \frac{1}{(1-\rho)} .$$

(To compensate for the particular sequences - ξ - ξ' - of length $2\ell+3$ involving two structures of length ℓ, which were counted as only one occurrence, we can determine their approximate probability, using the same method, and add that to the value just determined. However for the values of ρ of interest here this would be an extremely small correction and we will not use it.)

Q.E.D.

Corollary to Lemma 1. Free elements occur with density

$$D(1) > (1-\rho)^2 / (1+\rho-\rho^2)$$

under a uniform random distribution.

Lemma 2. The probability that the EXCHANGE operator will disrupt a structure with b weak links (by exchanging part of it with the surround) is

$$P_D = m_1[b + \frac{2}{m_2} + (1-m_2)^b(2b - \frac{2}{m_2})] \cong (>)\ (5b - 2b^2m_2)m_1 \qquad \text{for } m_2b < 1$$
$$\cong 5bm_1 \qquad \text{for } m_2b < 2\ .$$

proof:

Assume the outer limit of the EXCHANGE operator is c weak links from the pivot. Then for c>b there are 2c+b possibilities for the EXCHANGE operator to affect the structure,

c=2

b=3

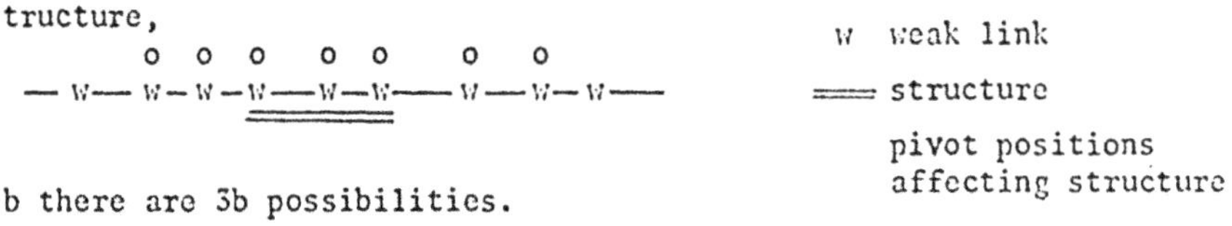

and for c = b there are 3b possibilities.

c=2

b=1

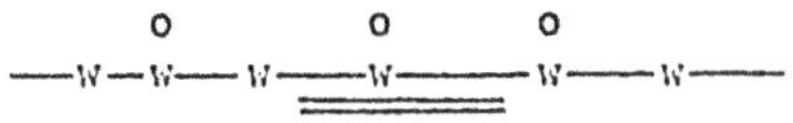

Thus, summing over all the possibilities for c, we get

$$P_D = \Sigma_{c=1}^{b-1}\, m_1(2c+b)\ \text{Prob\{outer limit} = c\} + \Sigma_{c=b}^{\infty}\, m_1(3b)\ \text{Prob\{outer limit} = c\}$$

But $\quad$ Prob{outer limit = c} $\overset{df.}{=} p_c = m_2(1-m_2)^{c-1}$

$$\text{Whence}\quad P_D = \Sigma_{c=1}^{b-1}\, m_1 2cm_2(1-m_2)^{c-1} + \Sigma_{c=1}^{b-1}\, m_1 bm_2(1-m_2)^{c-1} + \Sigma_{c=b}^{\infty}\, m_1 3bm_2(1-m_2)^{c-1}\ .$$

Each of these series can be summed using the equation

$$\Sigma_{j=0}^{\infty}\, r^j = \frac{1}{1-r} \qquad \text{for } 0 < r < 1,$$

and its derivative to give

$$P_D = 2m_1m_2\left[\frac{1-(1-m_2)^b}{m_2^2}\right] + bm_1m_2\left[\frac{1-(1-m_2)^b}{m_2}\right] + 3bm_1m_2\left[\frac{(1-m_2)^b}{m_2}\right]$$

$$= m_1[b + \frac{2}{m_2} + (1-m_2)^b(2b - \frac{2}{m_2})]\ .$$

The approximation follows from the fact that

$$(1-m_2)^b > 1-bm_2 \qquad \text{for } m_2 < 1\ .$$

Q.E.D.

Lemma 3. The probability that the EXCHANGE operator extends a given structure by adding a suffix structure at the immediate right is

$$P_E = 3m_1\, \rho$$

proof:

A suffix structure can be added only if the pivot or one of the outer limits is at the terminating weak link of the structure (its right end). The probability that the pivot falls there is simply m_1 while the probability that an outer limit falls there is

$$\Sigma_{h=1}^{\infty} \Pr\{\text{pivot is h units away and outer limit} = h\}$$
$$= m_1 \Sigma_{h=1}^{\infty} (1-m_2)^{h-1} m_2$$

Since there are two other outer limits, we have

$$P_E = [m_1 + 2m_1 m_2 \ \Sigma_{h=1}^{\infty} (1-m_2)^{h-1}]\rho$$
$$= 3m_1 \rho$$

where ρ is the probability that the new content of the cell at the right end is some (non-null) element.

Q.E.D.

Lemma 4. The expected number of interior weak links in a structure of length ℓ is $(\ell-1)/(\ +1)$ at equilibrium under the primitive operators (assuming effects of emergent operators negligible).

proof:

Let there be $M_S(t)$ strong links at time t, and $M_W(t)$ weak links. We can expect $rM_S(t)$ strong links to become weak and $\lambda r M_W(t)$ weak links to become strong. Thus, we can expect

$$M_S(t+1) = M_S(t) - rM_S(t) + \lambda r M_W(t).$$

At equilibrium we expect

$$M_S(t+1) = M_S(t).$$

Solving these two equations, we get the expected ratio of strong to weak links at equilibrium,

$$M_{S,\text{equil.}}/M_{W,\text{equil.}} = \lambda$$

Thus of the $(\ell-1)$ links in the structure, on the average $\lambda/(\lambda+1)$ will be strong and $1/(\lambda+1)$ will be weak.

Q.E.D.

Corollary to Lemma 4. The expected lifespan of a structure of length λ exceeds $[5m_1(\frac{\ell-1}{\lambda+1} + 2)]^{-1}$ at equilibrium.

proof:

An average string of length ℓ will have $(\ell-1)/(\lambda+1)$ interior weak links and a weak link at either end. Using Lemma 2 we see that at any given time-step the EXCHANGE operator will affect any one of these $((\ell-1)/(\lambda+1))+2$ links with probability exceeding $5m_1$. At worst each of these links will remain weak until one of them is affected. The expected time for this is

$$[5m_1(\frac{\ell-1}{\lambda+1} + 2)]^{-1} \qquad \text{Q.E.D.}$$

Theorem 1. Under the primitive operators, structures will be persistently distributed according to a uniform random distribution (as long as the effect of emergent operators is negligible) if

$$\lambda \doteq \frac{5(1-\rho)}{3\rho^2} - 1 \quad .$$

(This is a stable equilibruim under these conditions, so that the α-universe will go to it, regardless of the initial distribution over structures, so long as the emergent operators have neglible effect).

proof:

Let $M(t)$ be the expected density (expectation of the distribution) of structures at time t, and let $X(t)$ be the average number of weak links per structure (i.e. weak links interior to structures).

Then, using Lemma 2, we can expect approximately $5m_1X(t)$ structures to be disrupted at time t, yielding an increase of one structure for each disruption. Similarly, by Lemma 3, we can expect $3m_1\rho M(t)$ structures to be concatenated, reducing $M(t)$ accordingly. Thus

$$M(t+1) = M(t) + 5m_1X(t) - 3m_1\rho M(t)$$

At equilibrium $M(t+1) = M(t)$, yielding $5m_1X(t) = 3m_1\rho M(t)$. Under a uniform random distribution the expected density of structures of length ℓ is approximately $(1-\rho)^2\rho^\ell$ by Lemma 1. Thus

$$M_{random} = \Sigma_{\ell=1}^{\infty}(1-\rho)^2\rho^\ell = (1-\rho)\rho \ .$$

The average number of links (the length minus one) in a structure is then

$$L' = [\Sigma_{\ell=1}^{\infty}(\ell-1)(1-\rho)^2\rho^\ell] = \rho^2$$

Hence, by Lemma 4, $X_{random} = \frac{\rho^2}{\lambda+1}$

To maintain this uniform random distribution at equilibrium, then,

$$5m_1X_{random} = 3m_1\rho M_{random} \quad \text{or } \lambda \cong \frac{5(1-\rho)}{3\rho^2} - 1 \ .$$

Q.E.D.

Theorem 2. The probability that a given structure A of length ℓ will be formed by the EXCHANGE operator on a given time-step in a region of R cells is approximately

$$\left(\frac{5m_1}{\lambda+1}\right)(1-\rho)^2\rho_c^{\ \ell}(\ell-1)R,$$

where all elements other than the null element are assumed equilikely, with proportion ρ_c, and emergent operators are assumed to have neglible effect.

proof:

A can be divided into two fragments A_1 and A_2, of length ℓ_1 and ℓ_2 respectively, in $\ell-1$ ways.

The EXCHANGE operator will produce a fragment A_1 from some longer structure A' with probability

$$(5m_1)\left(\frac{1}{\lambda+1}\right)(1-\rho)\rho_c^{\ \ell_1},$$

where the factor $5m_1$ comes from Lemma 2, $1/(\lambda+1)$ is the probability that a weak link occurs at precisely the distance ℓ_1 from the left end of A', and $(1-\rho)\rho_c^{\ \ell_1}$ is the probability of occurrence of a structure with prefix A_1 (following Lemma 1).

The probability that the fragment brought into conjunction with A_1 by the EXCHANGE operator is A_2 is given by

$$(1-\rho)\rho_c^{\ \ell_2}$$

(again following Lemma 1).

Hence the EXCHANGE operator will produce A as a conjunction of particular fragments A_1 and A_2 with probability (i.e. density in the universe).

$$\left(\frac{5m_1}{\lambda+1}\right)(1-\rho)^2\rho_c^{\ \ell_1}\rho_c^{\ \ell_2}, \quad \ell_1+\ell_2 = \ell \ .$$

Since there are $(\ell-1)$ fragment pairs (A_1,A_2) yielding A and R cells in the region we get

$$\frac{5m_1}{\lambda+1}\ (1-\rho)^2\rho_c^{\ \ell}(\rho-1)R$$

Q.E.D.

4. EXPECTED TIME UNTIL THE EMERGENCE OF A SELF-REPLICATING SYSTEM.

In any study of the spontaneous emergence of self-replicating systems, it is the expected time until their emergence, not the fact that they will emerge, which is critical. The theorems just proved enable good estimates of these times in α-universes. In these calculations Theorem 1 plays a critical role, particularly in assuring that the primitive operators are unbiased with respect to the emergence of any structure.

As an example of a self-replicating system in the particular -universe presented, consider

$$0{:}\alpha(N_0N_0N_1) \qquad \gamma(0{:}\alpha(N_0N_0N_1))$$
$$0{:}\alpha(N_0N_1N_1) \qquad \gamma(0{:}\alpha(N_0N_1N_1))$$
$$1{:}\alpha(N_0N_0N_1) \qquad \gamma(1{:}\alpha(N_0N_0N_1))$$
$$1{:}\alpha(N_0N_1N_1) \qquad \gamma(1{:}\alpha(N_0N_1N_1))$$

where $\alpha(\cdot)$ designates the anticode of $(\cdot)$, e.g. $\alpha(N_0N_0N_1) = 001$, and $\gamma(\cdot)$ designates the code of $(\cdot)$, e.g. $\gamma(0{:}\alpha(N_0N_0N_1)) = N_0N_0N_1N_0N_0N_0N_0N_0N_0N_1$. The structures in the left columns are emergent operators; those in the right column are strings in the N-alphabet, or codon <u>strings</u>, serving as repositories of information.

Note that the argument of the first emergent operator, $0{:}\alpha(N_0N_0N_1)$ is satisfied by either of the first two codon strings, since $\gamma(0{:}\alpha(N_0N_0N_1))$ has the prefix $N_0N_0N_1$ as does $\gamma(0{:}\alpha(N_0N_1N_1))$. Thus, the operator $0{:}\alpha(N_0N_0N_1)$ will produce copies of both of these codon strings, in proportions depending upon their concentrations. (That is, according to its definition, the COPY operator selects its operand at random from amongst those structures satisfying its argument). In effect we have an instance of the second Ω-property, Ω.2. If the codon strings are at a high density relative to other structures with the same prefix, the COPY operator will exhibit a high efficiency in producing copies of these codon strings. Otherwise much of its effort will be wasted in producing copies of irrelevant structures.

As we will see shortly, the "equilibrium density" (under the uniform random distribution) of structures with any given 3-letter prefix is low enough that very few irrelevant structures will interfere with the copying procedure. This is <u>not</u> true for 2-letter prefixes; on the other hand, a 4-letter prefix is even more specific, but the increase in efficiency, vis-a-vis "equilibrium densities", is small. These conditions dictate the use of 3-letter arguments for the emergent operators in this self-replicating system.

In the same way, the second of the emergent operators copies the last two codon strings. The third emergent operator decodes both of the first two codon strings, producing new copies of the first two emergent operators; the fourth emergent operator decodes both of the second two codon strings, producing new copies of the second two emergent operators. Thus the eight structures interact in such a way as to produce replicas of all eight structures. This process of replication will continue until it exhausts the supply of some free element required by the copying or decoding process (the "resource limit"). The structures will then maintain the (high) densities they have achieved, by using "breakdown products" produced by the primitive operators. Any subsequent change which makes for more efficient replication will cause a "migration" of elements from the given self-replicating system to the more efficient one. In this way Darwinian selection, imposed by a competition for limited resources, begins.

At this point we would like to determine the expected time until such a combination of structures emerges, the emergence time. To do this we must first determine the rate at which new structures (of a given type) are generated in the α-universe (see Theorem 2). Having done this we can determine the number of new n-tuples (octuples in the above case) tested per time-step, and from this the expected time to reach a particular n-tuple.

Theorem 3. Given j particular structures having only strong bonds internally, of combined length ℓ, the emergence time is well approximated by

$$[(\ell-j)\ r(\tfrac{\lambda}{\lambda+1})^{\ell-j}\ (1-\rho)^{2j}\ \rho_e^{\ell}R^j]^{-1} \quad \text{for } \lambda r \ll 5m_1 \text{ and } r \ll 1.$$

proof:

(I will leave the generalization to j>2 structures to the reader.)

Let there be M'_ℓ new structures (generated by EXCHANGE) and M_ℓ total structures, of length ℓ, expected in R cells at any given time. Consider two structures A and B of lengths a and b respectively. Then on any given time-step there will $M'_aM_b + M'_bM_a - M'_aM'_b$ new pairs of structures. (If ℓ is large enough we can ignore the possibility of a given structure being generated repeatedly). Whence the expected time to the emergence of A with B is just

$$(M'_aM_b + M'_bM_a - M'_aM'_a)^{-1}\ .$$

Using Theorem 2 we see that $M'_\ell = \left(\frac{5m_1}{\lambda+1}\right)(1-\rho)^2\rho_e^{\ell}(\ell-1)R$ if we are interested in a particular structure of length ℓ, without specifying what bonds are involved. If we desire the structure to involve only strong bonds, then it must be generated from two fragments involving only strong bonds and the weak bond at the point of juncture must revert to a strong bond before the structure is disrupted again at that point (ignoring the possibility of disruption at one of the extant strong bonds). The probability of disruption at a weak bond, on any given time-step, is $5m_1$ (see Lemma 2); the probability of the weak bond becoming strong is λr; thus $\lambda r/(5m_1+\lambda r)$ gives the probability of the weak bond becoming strong before it is disrupted. The requirement that the other $(\ell-2)$ bonds (in the two fragments) the strong is met with probability $(\lambda/\lambda+1)^{\ell-2}$ (see Lemma 4). Thus for a structure of length ℓ involving only strong bonds

$$M'_\ell = \frac{5m_1}{(\lambda+1)}\ \frac{\lambda r}{5m_1+\lambda r}\left(\frac{\lambda}{\lambda+1}\right)^{\ell-2}(1-\rho)^2\rho_e^{\ell}(\ell-1)R.$$

$$\cong r\left(\frac{\lambda}{\lambda+1}\right)^{\ell-1}(1-\rho)^2\rho_e^{\ell}(\ell-1)R \qquad \text{for } r \ll 5m_1$$

Using Lemma 1 and an argument similar to the one above we have

$$M_\ell = \left(\frac{\lambda}{\lambda+1}\right)^{\ell-1}(1-\rho)^2\rho_e^{\ell}R.$$

Bringing all of this together we have

$$M'_aM_b = (a-1)r\left(\frac{\lambda}{\lambda+1}\right)^{a+b-2}(1-\rho)^4\rho_e^{a+b}R^2\ ,$$

$$M'_aM_b = \frac{(b-1)}{(a-1)}M'_aM_b\ ,$$

so that

$$(M'_aM_b + M'_bM_a - M'_aM'_b)^{-1} \cong [(a+b-2)r\left(\frac{\lambda}{\lambda+1}\right)^{a+b-2}(1-\rho)^4\rho_e^{a+b}R^2]^{-1} \qquad \text{for } r \ll 1,$$

(using the fact that $M'_aM'_b \ll M'_aM_b$ for $r \ll 1$).

Q.E.D.

If we set ρ to 1/2 and, to maintain a random distribution over structures (Theorem 1),

$$\lambda = \frac{5(1-\rho)}{3\rho^2} - 1 = \frac{7}{3},$$

then treating all elements as equilikely (i.e. $\rho_e = \frac{1-\rho}{5} = 10^{-1}$) we obtain an expected waiting time for the self-replicating octuple (ℓ = 60) of

$$[52r(.7)^{52}\cdot(1/2)^{16}\cdot 10^{-60}\cdot R^8]^{-1} = (1.4 \times 10^{71})r^{-1}R^{-8}$$

Setting R to 10^4 cells and $r = 10^{-4}$ (a value which will prove to be significant later), we get a waiting time of 1.4×10^{43} time steps. This is such a large number that, for all practical purposes, we can reject the possibility of spontaneous emergence, if indeed the system must emerge in one fell swoop.

The calculation above closely parallels that in the usual argument <u>against</u> spontaneous emergence, however it ignores much that is essential if we are to make a realistic estimate. The argument is like that which uses the mammalian eye as an argument against the Darwinian interpretation, claiming that the eye must emerge in one step, because there would be no intermediate forms with selective advantage. In fact there are advantageous intermediate forms leading to the eye (as Darwin carefully shows). Similarly, there are steps to the self-replicating system which build upon each other and vastly reduce the waiting time.

To begin with there is a different version of this self-replicating system which relies on the EXCHANGE operator for assembly of some of the components:

$$0{:}\alpha(N_1N_0N_0) \qquad \gamma({:}\alpha(N_1N_0N_0)) = N_1N_0N_0N_1N_0N_0N_0N_0$$

$$1{:}\alpha(N_1N_0N_0)$$

The prefix $N_1N_0N_0$ of $\gamma({:}\alpha(N_1N_0N_0))$ is just that specified by the two emergent operators at the left. Thus the first operator will make additional copies of the codon string while the second operator will decode it to produce a high concentration of $:\alpha(N_1N_0N_0)$ = :100. When :100 reaches a high density the EXCHANGE operator will join 0s (and 1s) to a fair proportion of the copies, to form 0:100 (and 1:100), thus completing the cycle of replication.

Actually, if the density of the codon string is great enough, the two emergent operators will be efficient with only a 2- letter prefix, i.e.

$$0{:}\alpha(N_1N_0) \qquad 1{:}\alpha(N_1N_0)$$

and the codon string can be shortened, accordingly, to

$$\gamma({:}\alpha(N_1N_0)) = N_1N_0N_0N_1N_0N_0 .$$

(This point will soon be established quantitatively via Theorem 4).

Finally, if there is a high density of $({:}\alpha(N_1N_0))$ and it has a long half-life (low probability of disruption by the EXCHANGE operator) then $1{:}\alpha(N_1N_0)$ = 1:10 has a reasonable probability of being formed spontaneously by the EXCHANGE operator while the density of $({:}\alpha(N_1N_0)$ is still high (see Theorem 4). Thus, 1:10 is not required for the "start-up" of the system, under these conditions.

Coupling all these facts, we see that the pair

$$0{:}\alpha(N_1N_0N_0) \qquad \gamma({:}\alpha(N_1N_0))$$

in a region of R cells is a sufficient "seed" for the self-replicating system. (In the "seed", the operator $0{:}\alpha(N_1N_0N_0)$ needs the 3-letter prefix to assure that it will preferentially attach to the single copy of $\gamma({:}\alpha(N_1N_0))$.

To establish the effectiveness of this "seed" system we need a theorem which details the rate of increase (or decrease) of $\gamma({:}\alpha(N_1N_0))$ and :10 under the effects of the various operators in the α-universe. Let $M_R(A)$ be the number of copies of structure A in some region of R cells and let $\Delta_T M_R(A)$ be the rate of change of $M_R(A)$ over a period of time T.

Theorem 4. For $T << 1/r$: $\Delta_T M_R(\alpha(:10))$

$$\cong [-\frac{5r}{2} M_R(\gamma(:10)) + (3m_1)(1-\rho)\rho_e(1/2)\, M_R(:10) + \left(\frac{5m_1}{\lambda+1}\right)\left(\frac{\lambda}{\lambda+1}\right)^2 (1-\rho)^2\rho_e{}^4(3R)\beta\,]T$$

where $\beta = \left(\frac{1}{5m_1}\right)\left(\frac{M_R(\gamma(:10))}{6(1-\rho)^2\rho_e{}^2+10M_R(\gamma(:10))}\right)$

$=^{\text{df.}}$ the number of decodings of the $\gamma(:10)$ produced by 0:10 before it is disrupted.

$\Delta_T M_R(:10)$

$$\cong [-\frac{2r}{2} M_R(:10) + (3m_1)(1-\rho)\rho_e(1/2)\beta' M_R(:10) + \left(\frac{5m_1}{\lambda+1}\right)\left(\frac{\lambda}{\lambda+1}\right)^2 (1-\rho)^2\rho_e{}^4(3R)\beta'\,]T$$

where $\beta' = \left(\frac{1}{5m_1}\right)\left(\frac{M_R(\gamma(:10)}{5(1-\rho)^2\rho_e{}^2+7M_R(\gamma(:10))}\right)$

$=^{\text{df.}}$ the number of decodings of $\gamma(:10)$ produced by 1:10 before it is disrupted.

proof:

Three factors enter into changes in the number of copies of $\gamma(:10)$:

(f1) "Decay" of extant copies of $\gamma(:10)$ because a strong bond therein changes to a weak bond (by BOND MODIFICATION) and is disrupted by EXCHANGE.
(f2) New copies of $\gamma(:10)$ produced by copies of 0:10 resulting from attachment of 0 to :10 by the EXCHANGE operator.
(f3) New copies of $\gamma(:10)$ produced by copies of 0:10 resulting from random actions of the EXCHANGE operator.

That is, symbolically, $\Delta_T M_R(\gamma(:10)) = [-(f1) + (f2) + f3)]T$.

Factor (f1) is determined by noting that the time for 1/2 the strong bonds at a given position to be transformed to weak bonds is just $1/r$. There are 5 internal links in $\gamma(:10)$ thus the "decay" rate per time-step is $5r/2$.

Factor (f2) is determined by using Lemma 3 together with Lemma 1 to determine the probability that the prefix 0 is added (by EXCHANGE) to :10 on a given time-step: $(3m_1)(1-\rho)\rho_e$. One half of these will be terminated by a - at the right of 0:10. For each copy of :10 so transformed we must determine how many of the structures it copies, before it is disrupted, will be $\gamma(:10)$. This depends upon the concentration of $\gamma(:10)$ relative to all other structures with prefix N_1N_0 and the time it takes to copy the structures. Taking into account that almost all other structures with prefix N_1N_0 will in fact be the 2-symbol structure N_1N_0, and noting that it takes $\ell+4$ steps to copy a structure of length ℓ, we get

$$\frac{M_R(\gamma(:10))}{6(1-\rho)^2\rho e^2+10M_R(\gamma(:10))}$$

using Lemma 1 (and assuming that we are operating in a random distribution of structures). 0:10 has one weak bond so that it can be expected to operate $1/5m_1$ time steps before it is disrupted, yielding

$$\beta = \left(\frac{1}{5m_1}\right)\left(\frac{M_R(\gamma(:10))}{6(1-\rho)^2\rho_e{}^2+10M_R(\gamma(:10))}\right) .$$

Thus factor (f2) is $(3m_1)(1-\rho)\rho_e(1/2)\beta M_R(\gamma(:10))$.

Factor (f3) depends upon the "spontaneous" production of 0:10 by the EXCHANGE operator, rather than construction from :10. Thus, if we require 0:10 to have only one weak link, we get (using Theorem 2) the probab[illegible]

$$\left(\frac{5m_1}{\lambda+1}\right)\left(\frac{\lambda}{\lambda+1}\right)^2 (1-\rho)^2 \rho_e^{\,4} (3R) \quad .$$

Otherwise, the calculation exactly parallels that for factor (f2), yielding

$$\left(\frac{5m_1}{\lambda+1}\right)\left(\frac{\lambda}{\lambda+1}\right)^2 (1-\rho)^2 \rho_e^{\,4} (3R)\beta \quad .$$

Substituting the expressions for (f1), (f2), and (f3), just derived, in [-(f1)+(f2)+(f3)]·T yields the expression for $\Delta_T M_R(\gamma(:10))$ in the statement of the theorem.

The derivation of the expression for $\Delta_T M_R(:10)$ exactly parallels the above derivation, the only differences arising because the decoding time is $(\ell/2)+4$, and the half-life of :10 is 2r/2 since there are only two links in :10.

Q.E.D.

Using $\rho = 1/2$, $\lambda = 7/3$, $\rho_e = 10^{-1}$, and $R = 10^4$, as we did earlier, and noting that m_1 actually cancels out of the relations for $\Delta_T M_R(\gamma(:10))$ and $\Delta_T M_R(:10)$, we obtain

$$\frac{\Delta_T M_R(\gamma(:10))}{\beta} = -\frac{5r}{2\beta} M_R(\gamma(:10)) + 7.5\cdot 10^{-4} M_R(:10) + 5.51\cdot 10^{-3}$$

and
$$\frac{\Delta_T M_R(:10)}{\beta' M_R(:10)} = -\frac{r}{\beta'} + 7.5\cdot 10^{-4} + \frac{5.51\cdot 10^{-3}}{M_R(:10)}$$

Noting that β and $\beta M_R(:10)$ are always > 0 we quickly obtain conditions for $\Delta_T M_R(\gamma(:10))>0$ and $\Delta_T M_R(:10)>0$ simultaneously, namely $(1.66r\cdot 10^3)M_R(\gamma(:10)) < M_R(:10) < \dfrac{M_R(\gamma(:10))}{1.26r\cdot 10^3}$ for $r < 2\cdot 10^{-4}$.

Thus, if $r = 10^{-4}$, we see that both $\gamma(:10)$ and :10 will increase simultaneously if $\dfrac{M_R(\gamma(:10))}{6} < M_R(:10) < 8M_R(\gamma(:10))$.

In the "seed" $[0{:}\alpha(N_1N_0N_0),\ \gamma(:\alpha(N_1N_0))]$ the copy operator $0{:}\alpha(N_1N_0N_0)$ will produce about one copy of $\gamma(:\alpha(N_1N_0))$ every 10 time-steps over its half-life (which is 4·r). Hence, for $r = 10^{-4}$, $M_R(\gamma(:10))$ will soon increase to a concentration where an operator 1:10 formed spontaneously by the EXCHANGE operator will decode $\gamma(:10)$ to provide $M_R(:10)>1$. By Theorem 2 we can expect a "spontaneous" 1:10 every

$$\left[\left(\frac{5m_1}{\lambda+1}\right)(1-\rho)^2\rho_e^{\,4}3R\right]^{-1} \cong 115 \qquad \text{time-steps for } m_1 = 10^{-2} \ .$$

Thus we soon reach a condition where both $\gamma(:10)$ and :10 increase to the limits set by resources (free elements) and their own breakdown rates. At that point the self-replicating system establishes itself, operating according to the reaction diagram given below. In other words, once the "seed" occurs the self-replicating system develops and expands through out the space.

Using theorem 3, with $\ell=11$, we can quickly calculate the waiting time for the spontaneous emergence of the "seed", using the same parameter values as for earlier self-replicating systems. We obtain $[(9)\cdot 10^{-4}\cdot(.7)^9\cdot(1/2)^4\cdot 10^{-11}\cdot(10^4)^2]^{-1} = 4.4\cdot 10^8$ time-steps as compared to the earlier $1.4\cdot 10^{43}$ time-steps. That is, if we take a time-step to be 1 millisecond, then the "seed" would emerge spontaneously in about 125 hours while the full-blown system would emerge (in one step) only after a time in excess of $4\cdot 10^{30}$ centuries!

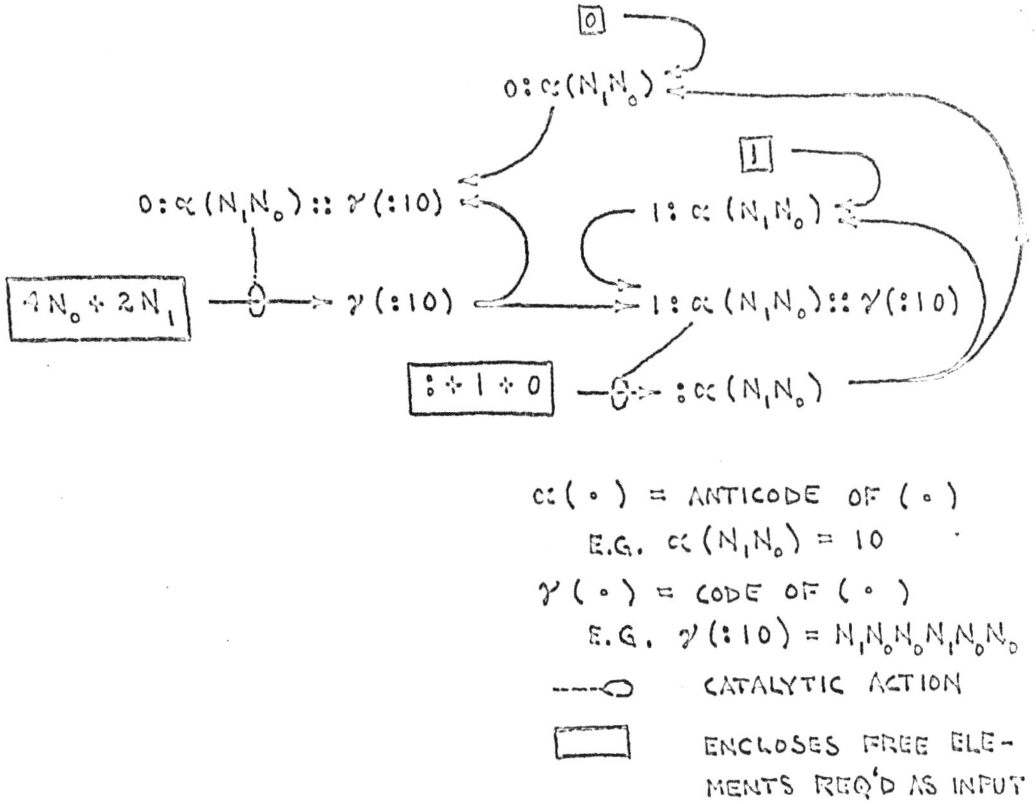

Thus, by taking into account the steps by which a self-replicating system could arise, rather than requiring it to emerge en toto, we go from impossibility to times compatible with computer simulation. The next section will explore the implications of these results.

5. COMMENTARY

At the center of this study is a formally defined artificial universe which begins in "chaos" and, after a lapse of time, contains "life" in the sense of self-replicating systems undergoing heritable adaptations. The requirement that the primitive ("kinetic") operators preserve the "chaos" (the uniform random distribution over elements) keeps the universe from any particular predisposition toward "living systems". Only emergent (catalytic-like) operators, spontaneously generated by the random mixing, modify the "chaos". The time involved is short enough to be compatible with computer simulation.

Real living systems, even conjectured primordial systems, are considerably more complex than the systems set forth here. This would suggest a large increase in emergence times. Counterbalancing this are the very high "mixing rates" (the counterparts of m_1 and m_2 in the α-universe) and the great numbers of atoms involved (the counterpart of $_\rho R$) in natural domains of interactions. The carbon atoms in a cubic centimeter of 10^{-6} molar solution will undergo in excess of 10^{30} collisions (potential interactions) per second at room temperature. This would be like investigating a region R in the α-universe of 10^{16} cells with a time-step equal to 10^{-12} second!

With this in mind, it seems worthwhile to investigate, by both analysis and simulation, more complex α-universes reflecting more of the known properties of biomolecular solutions. There are several modifications yielding greater realism which do not much alter the theorems and lines of approach laid out here; generally they can be incorporated without much difficulty when they are needed for particular lines of investigation:

1) Underlying cellular spaces of dimension 2 or 3. The major change is redefinition of the EXCHANGE operator, providing more straightforward random walks of elements and structures. It is even possible to introduce a simple ENERGY EXCHANGE operator, which provides an energy distribution over structures, of the Maxwell - Boltzmann type, and an appropriate equipartition of energy principle. The BOND-MODIFICATION operator can be redefined to be sensitive to the energies so defined.

2) Element-specific bonds. The range of bonds can be increased to a set for each element of interest, with a corresponding increase in the number of parameters of type λ and r. In combination with the ENERGY EXCHANGE operator of (1), this provides a wide range of stabilities for structures, with a corresponding increase in varieties of "grist" for the "selection mill".

3) All emergent operators required to be strictly catalytic. The emergent operators can be restricted to joining (or disjoining) structures which are brought into contact with them by the EXCHANGE operator. "Active sites" (e.g. sites requiring certain prefixes or suffixes on the structures to be acted upon) would be defined by bond specificities in the elements involved, while the positioning of the "active sites" could be given a first order approximation by requiring chains of elements of specific lengths and types between the sites.

4) Compartmentation. Natural coherence (cf. coazervates) and the effects of membranes can be approximated by making the operators (particularly the EXCHANGE operators) sensitive to certain "punctuations" in the cellular space. In the particular α-universe detailed earlier, this could be accomplished, for instance, by prohibiting the EXCHANGE operator from producing exchanges across any strongly-linked pair of :s. Higher level operators could then be introduced to treat a set of cells bounded by double :s as a compartment. For example, one could define a DEPOSIT operator which moves a particular operand from one side of the double : to the other (selective transport). A FISSION operator could introduce a double : inside a compartment whenever its element density exceeds a threshold (cf. fission of coazervates under surface tension), etc. By making the level of punctuation an operator is sensitive to a part of its specification (for emergent operators) - mimicing active transport, etc. - it is possible to observe the emergence of functional hierarchies.

The α-universe detailed earlier is only a proof-of-principle model, but with additions such as the foregoing, it should be possible to inject a much more realistic "chemistry". At that point models such as those of Oparin 1936, Bernal 1951, and Eigen 1971, could be investigated formally and some parts of them, at least, could be simulated.

Three lines of investigation seem well within reach, and quite profitable, at this point in time:

1) Investigation of the evolution of codons as a function of different types of element-specific bonds (instead of predetermining the codes as in the proof-of-principle α-universe). It should be possible to characterize a range of conditions leading to the emergence of codons.

2) Investigation of the relation between "compartmentation", functional hierarchies, and stability in emergent self-replicating systems. It should be possible to characterize factors leading to increasing specialization of function even at this primitive level.

3) Investigation of the effect of various proportions of elements upon the succession of self-replicating systems. It should be possible to demonstrate

a kind of speciation wherein, once easily available elements and structures are used up by one emergent self-replication system, other self-replicating systems emerge to fill unused niches (i.e. to utilize less simply exploited elements).

Overall, there is much more to be learned about the relation between intrinsic parallelism (see the discussion of + schemata at the end of section 1) and the speed with which self-replicating systems emerge and evolve.

REFERENCES

Bernal, J.D. (1951). The Physical Basis of Life, (Routledge and Kegan Paul, London).

Burks, A.W., ed. (1970). Essays on Cellular Automata. (University of Illinois Press, Urbana).

Eigen, M. (1971). Self organization of matter and the evolution of biological macromolecules, Naturwissenschaften, 10.

Feller, W. (1950). Probability Theory and Its Applications. (Wiley, New York).

Lindenmayer, A. (1968). Mathematical models for cellular interactions in development, J. Theore. Biol., 18.

Holland, J.H. (1975). Adaptation in Natural and Artificial Systems. (University of Michigan Press).

Moorhead, P.S., and Kaplan, M.M., eds. (1967). Mathematical Challenges to the Neo-Darwinian Interpretation of Evolution. (Wistar Inst. Press, Philadelphia).

Oparin, A.I. (1936). Origin of Life, 1938 transl. (Macmillan, New York).

LOGIC OF COMPUTERS GROUP

NEW PERSPECTIVES IN NONLINEARITY
OR
WHAT TO DO WHEN THE WHOLE IS MORE THAN THE SUM OF ITS PARTS

John H. Holland

Department of Computer and Communication Sciences
2080 Frieze Building
The University of Michigan
Ann Arbor, Michigan 48109

New Perspectives in Nonlinearity
or
What to Do When the Whole Is More Than the Sum of Its Parts

John H. Holland
The University of Michigan

This paper presents a theory of algorithms designed to optimize highly interactive systems (multi-dimensional, multi-peak, nonlinear functions). Two applications are discussed: one concerns cognitive systems capable of learning and generalization, and one concerns calculations dealing with the "origin of life" from "organic soups". The algorithms are *intrinsically parallel* -- each function argument processed serves as a carrier for information about a tremendous number of regions (hyperplanes) in the function's domain. Each region is automatically ranked according to the estimated average value of the function over that region, and the rankings are compactly stored in the algorithm's data base (M ℓ-tuples store approximately $M \cdot 2^{\ell/2}$ rankings). Thus, the algorithm implicitly processes hyperplanes by manipulating function arguments.

New Perspectives in Nonlinearity
or
What to Do When the Whole Is More Than the Sum of Its Parts

John H. Holland

The University of Michigan

Before starting, I would like to beg the reader's indulgence. This paper is not a survey, or even a balanced account mentioning related work. It is simply an exposition of one approach to the problem. As such it contains only three references -- all to my own work! Mea culpa.

In each of the sciences there are pivotal problems which can be formalized as searches for the global optimum of a function f: $A \rightarrow R$, where A is an appropriately chosen domain of elements and R is the set of positive real numbers. In genetics, for example, A becomes the genotypes (combinations of genes) of interest and f assigns to each genotype its fitness; in economics (von Neumann's model), A becomes mixes of economic activities and f assigns to each mix its utility; in game playing, A becomes the possible strategies and f assigns to each strategy its minimax value; and so on. Characteristically, f involves a multitude of local optima ("false peaks") when the problems are interesting or useful. (Each of the above examples generates such an f). When local optima abound most of the tools of mathematics, which amount to "linearization" in one guise or another, become quite awkward, and the gradient ("hill-climbing") algorithms of the computer sciences perform little better than an exhaustive search of A. (Exhaustive searches are of staggering inefficiency when A is at all large, and interesting spaces of more than 10^{30} elements are typical -- chromosomes having 100 or more genes with 2 or more alleles (variants) per gene, games having 4 or more alternatives per move and requiring an average of 50 or more moves for completion, etc. An exhaustive search of a space of 10^{30} elements would require about 10 billion years at a rate of 10^{12} elements per second!) No technological "fix" will bring multiple peak problems within range if exhaustive search is the only option. The critical question is: Do there exist, useful alternatives for searching out the global optima of such nonlinear (multiple peak) functions?

In the simplest sense a nonlinear function f: $A \to R$ is difficult to optimize because the value of f cannot be determined by the superposition (summing) of functions defined on components of A. We cannot optimize the function by finding an optimal setting for each of its components -- the interaction of the components yields an action of the whole quite different from the sum of the actions of its parts. It is my intention here to lay down some general principles for tackling such problems. The first part of the paper is devoted to a theoretical discussion of these principles and the kinds of algorithm that result. In answer to the critical question, we will see that there do indeed exist algorithms which do better than either exhaustive search or "linearization". The last part of the paper discusses the application of the ideas developed to two broad problem areas. The first of these concerns the construction of models of cognition, specifically models capable of learning and generalization in realistic times. The second concerns a comparison of arguments for and against the emergence of life (self-replicating systems) from "organic soups" (homogeneous pre-biotic substrates).

I. Theory

An approach to this question of alternative search procedures involves three steps:

1) The "structure" of the elements in A must be *represented*. This makes possible investigation of "regularities" in the variation of f over parts of A; elements of A can be compared and credit provisionally apportioned to components associated with higher values of f. The difficulty of the search is determined by the number and reconditeness of these regularities in relation to the global optimum.

2) Realistic bounds must be established for the time required to attain the global optimum as a function of the search's difficulty. This allows a comparison of algorithms over various ranges of functions, {f}. An algorithm, to be of interest, must be *robust* over an interesting range of functions, in the sense of approaching the bounds over that range.

3) Robust algorithms must be constructively defined.

Representation of A is a necessary first step for consideration of constructive optimization procedures. We will assume that each element in A can be given a *unique* representation as an ℓ-tuple of attributes. (E.g., in genetics, a chromosome with ℓ genes would be represented by specifying for each gene the particular allele, or alternative form, of the gene carried by that chromosome. A real number x, say $0 \leqq x < 1$, can be represented to 20 bits accuracy by the 20 bit binary expansion of the number. And so on.) A set of attribute functions δ_i: $A \to V_i$, $i = 1,\ldots,\ell$, will be used to represent elements of A, where $A \epsilon A$ has attribute v just in case $\delta_i(A) = v \epsilon V_i$. (E.g., in genetics, V_i would be the set of alleles for the ith gene; for the binary expansion of a real number, $V_i = \{0,1\}$ for all i). We can think of the δ_i as detectors which pick out features or components of A. The representation of $A \epsilon A$ is given by the ℓ-tuple $(\delta_1(A), \delta_2(A), \ldots, \delta_\ell(A))$. In other words, A is uniquely characterized by the set of properties it possesses.

In all that follows we will be dealing only with representations, so from here onward A will be taken to *be* the set of representations.

In these terms f is *linear* (relative to the representation) when, for all $A \epsilon A$,

$$f(A) = \Sigma_{i=1}^{\ell} \delta_i(A) .$$

(For present purposes, broadening the definition to

$$f(A) = \Sigma_{i=1}^{\ell} c_i \delta_i(A)$$

adds no real generality, because this just amounts to a new representation in terms of the detectors $\delta_i' = c_i \delta_i$). When f does have a linear representation, the global optimum is simply obtained. Because the coordinates do not interact they can be optimized independently of one another and then summed to yield the global optimum. (The whole *is* equal to the sum of its parts). Unfortunately most functions generated by large systems have no natural linear representation because the systems' components are highly interactive. "Almost all" such nonlinear functions have large numbers of peaks; they are the functions of interest here.

Given a representation, it is instructive to look at sets of elements which have one or more attributes in common. In order to name these subsets of A we will use a special symbol, "$\square$". "$\square$" indicates that we "don't care" what attribute occurs at a given position in the representing ℓ-tuple. Accordingly $(v, \square, \square, \ldots, \square)$ designates the subset of A consisting of all elements possessing attribute $v \epsilon V_1$ for detector δ_1. Equivalently $(v, \square, \square, \ldots, \square)$ consists of the set of all ℓ-tuples in A beginning with symbol v. Similarly $(\square, v', \square, v'', \square, \square, \ldots, \square)$ designates the subset of A consisting of all elements possessing attribute $v' \epsilon V_2$ *and* $v'' \epsilon V_4$ (and any attribute whatsoever for δ_1, $\delta_3, \delta_5, \delta_6, \delta_7, \ldots, \delta_\ell$). The subsets which can be so named will be called *schemata*. (They are hyperplanes in the representation space). In general the set of schemata is named by the product set

$$\Xi = \Pi_{i=1}^{\ell} \{V_i \cup \{\square\}\} .$$

$A \epsilon A$ belongs to the schema $\xi = (\Delta_1, \Delta_2, \ldots, \Delta_\ell) \epsilon \Xi$ if and only if (i) whenever $\Delta_j = v \epsilon V_j$, $\delta_j(A) = v$ and (ii) whenever $\Delta_j = \square$, $\delta_j(A)$ may have any value whatsoever. That is, whenever $\Delta_j = \square$ we don't care what attribute A has, but whenever Δ_j specifies some attribute $v \epsilon V_j$, A *must* have that attribute to belong to the subset $\xi = (\Delta_1, \Delta_2, \ldots, \Delta_\ell)$. It follows that the schema ξ consists exactly of those elements of A having the particular attributes specified by those $\Delta_i \neq \square$ in the ℓ-tuple naming ξ. (The reader is referred to Chapter 4 of reference [1] for an extended discussion of schemata).

We can illustrate the role of schemata in a familiar context by considering a function f of one variable x on the interval $0 \leqq x < 1$.

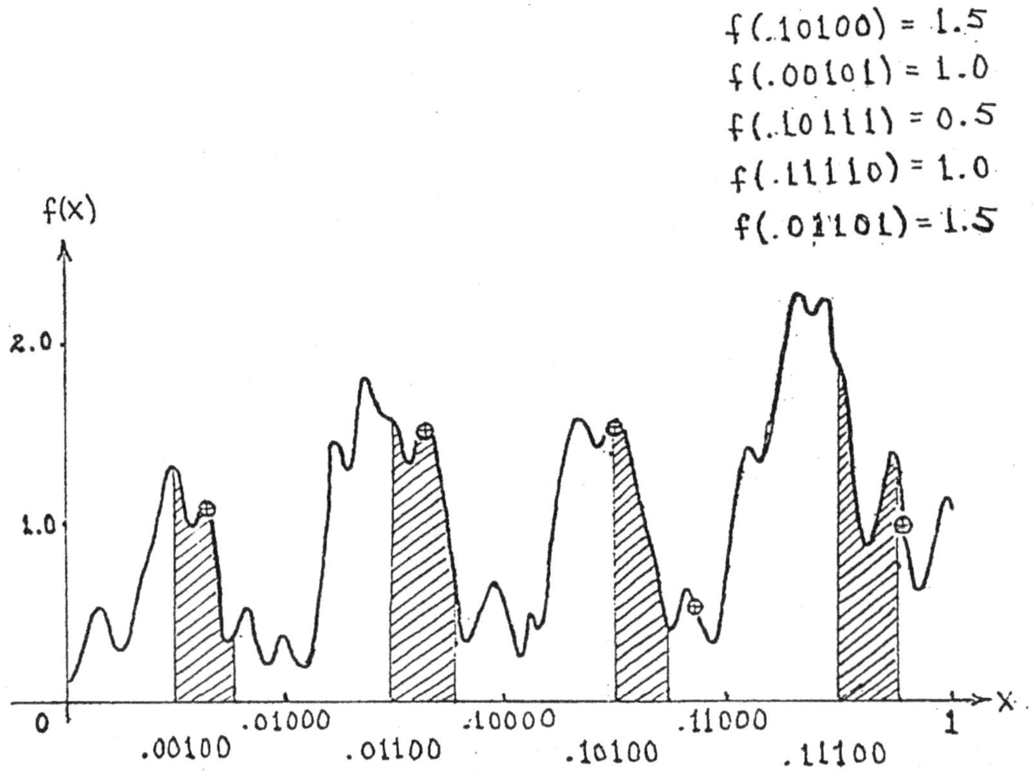

Here the domain A of the function has been given a binary representation to an accuracy of 5 bits. (Only 5 bits are used so that the representations will be short for the purposes of the illustration). Points belonging to the schema (□,□,1,0,□) fall in one of the four shaded strips. That is, the schema (□,□,1,0,□) is the union of the four intervals $1/8 \leqq x < 3/16$, $3/8 \leqq x < 7/16$, $5/8 \leqq x < 11/16$, and $7/8 \leqq x < 15/16$, as can be seen by making all possible substitutions at the three □'s.

Above the plot of f, the values of f for five points drawn at random from A are tabulated. If we think of these points as drawn according to some random distribution P over A, then A becomes a sample space and f becomes a random variable, each $A \in A$ occuring with probability P(A) and yielding value f(A). More importantly under distribution P each schema ξ is a random event and f restricted to ξ,

$f|\xi$, is a random variable with a well-defined expectation

$$f_\xi = [\Sigma_{A\varepsilon\xi} f(A)\cdot P(A)]/[\Sigma_{A\varepsilon\xi} P(A)] \ .$$

Any points drawn from the subset ξ constitute a legitimate sample of the event from which we can form an estimate $\hat{f}_\xi$ of the expectation f_ξ. Thus from the five points tabulated we can form a two point estimate of $f_{(0,\square,\square,\square,\square)}$

$$\hat{f}_{0\square\square\square\square} = (1.0 + 1.5)/2 = 1.25 \ .$$

(Here we abbreviate $(0,\square,\square,\square,\square)$ to $0\square\square\square\square$ and assume that the distribution is uniform -- all points are equiprobable. These conventions will hold throughout the example.) Similarly we can form a three point estimate of $f_{1\square\square\square\square}$,

$$\hat{f}_{1\square\square\square\square} = (1.5 + 0.5 + 1.0)/3 = 1.00 \ ,$$

and a three point estimate of $f_{\square\square 10\square}$,

$$\hat{f}_{\square\square 10\square} = (1.5 + 1.0 + 1.5)/3 = 1.33 \ .$$

Because of the higher estimated average for f over $\square\square 10\square$, we might consider modifying the distribution so that points are drawn more frequently from this schema than the other two. More of this later. For now, it is worth noting that *each* point drawn from A is a legitimate sample point of 2^ℓ distinct schemata. (Any schema defined by substituting $\square$'s for one or more of the ℓ attributes in the representation of A is a schema containing A. There are 2^ℓ ways of making such substitutions. Even if ℓ is only 20, we obtain information about more than one million schemata each time a point is evaluated!) If this tremendous amount of information can be accumulated and used to direct further sampling, we have a chance of quickly discovering and exploiting regularities in f.

To generalize the foregoing example, consider a "data base" of M elements $\{A_1,\ldots,A_M\}$ drawn from A, with the corresponding observed values of f, $\{f(A_1),\ldots,f(A_M)\}$. We would like to use this information to determine where to draw additional points from A in our search for the global optimum. Let us assume for the moment that the points $\{A_1,\ldots,A_M\}$ are more or less randomly scattered over A. Then one reasonable heuristic suggests that we draw additional points from "regions" in A where above-average values of f have been observed. Of course this heuristic is hopelessly vague without a prior definition of "region". Even with the definition, it is necessary to specify the precise manner in which the data base "biases" further point selections, if we are to evaluate the heuristic's usefulness. Let us see what can be done.

Earlier, using the given representation, we defined a rich class of subsets of A, the schemata. The above heuristic can be given a rigorous formulation in terms of schemata as follows. Let $P_\xi(t)$ be the probability of drawing a point (sampling) schema ξ at time t, let $\hat{f}_\xi(t)$ be the *observed* average value of f for the points drawn prior to time t from ξ, and let $\hat{f}(t)$ be the observed average value of f for all points drawn (from A) prior to t. Then, following the heuristic, a change ΔP_ξ in the probability of sampling ξ would be determined by the difference $\hat{f}_\xi(t) - \hat{f}(t)$. That is, if the average value of f over points drawn from ξ is greater than the average of f over all points drawn, P_ξ should be increased (and vice-versa). For example, the recursion

$$P_\xi(t + 1) = [\hat{f}_\xi(t)/\hat{f}(t)]P_\xi(t)$$

produces the suggested changes (while automatically assuring that the probabilities sum to 1 over sets of schemata which partition A). Specifically, this recursion yields

$$\begin{aligned}\Delta P_\xi &\overset{\text{df.}}{=} P_\xi(t + 1) - P_\xi(t)\\ &= [\hat{f}_\xi(t) - \hat{f}(t)]P_\xi(t)/\hat{f}(t)\end{aligned}$$

The heuristic so-specified can be supported rigorously along the following lines. Consider a finite set of random variables $\{\xi_1, \ldots, \xi_r\}$, each having an arbitrary distribution with a well-defined (but unknown) mean. At each instant $t = 1,2,3,\ldots$ we are allowed to choose (sample) one of the random variables, say ξ_j, and receive as "payoff" the value $f_j(t)$ observed (the value of f for the particular sample point drawn from the sample space underlying ξ_j). Our objective is to maximize the expected *sum* of the payoffs received after making a sequence of T choices. (That is, we want to use the information received from sampling the random variables to concentrate the samples on the random variable with the highest mean). This objective is attained if the proportion of trials allocated to ξ_j up to time T, $U_j(T)$, approximates

$$c \cdot \exp[\hat{f}_j(T) - \hat{f}(T)]$$

where c is a constant (based on the means and variances of the random variables). (Theorem 5.3 of reference [1] gives a precise statement of this result). The result holds for arbitrary random variables -- that is, arbitrary f's and distributions -- as long as the means and variances are defined.

It is easy to show that the proportions $U_j(T)$ will be produced (with probability 1, to arbitrary accuracy with increasing T) if the probability $P_j(t)$ of sampling ξ_j at time t changes according to the difference equation

$$\begin{aligned}\Delta P_j &\overset{\text{df.}}{=} P_j(t + 1) - P_j(t)\\ &= [\hat{f}_j(t) - \hat{f}(t)]P_j(t)/\hat{f}(t)\ .\end{aligned}$$

(See p. 137 of reference [1]). But this is exactly the rate of change of P_j specified by the heuristic, applied now to the random variables $\{\xi_j\}$ as counterparts of the schemata $\{\xi \subset A\}$.

Thus, if we can look upon the schemata $\{\xi\}$ as random variables and sample them with probabilities given by the algorithmic version of the heuristic

$$P_\xi(t+1) = \hat{f}_\xi(t)/\hat{f}(t)P_\xi(t) \quad ,$$

we will optimize the expected value of the *sum* of values of f. (This optimization is relative to any plan taking the same number of samples from the set of schemata $\{\xi\}$, regardless of the form of f; it is subject to the usual stochastic qualifications of "occurence with probability 1", etc.). This is *not* to say that the heuristic, as given, is the most efficient way of locating the *global optimum*. However, it cannot be too far off. Ultimately the expected sum can only be optimized if the point corresponding to the global optimum is frequently sampled. Moreover the exponential concentration of trials (the set of proportions U_j) makes it clear that this time cannot be long delayed (relative to the time it takes to build confidence in estimates of f_ξ). It is also true that for many purposes, optimization of the expected sum is preferable to optimization of the expected maximum. If the search will be a long one (say, of unknown length), it is often important that information be exploited as uncovered, so that the average performance of the algorithm (the average value of points sampled) not remain low for an indefinite period. This amounts to optimization of the expected sum.

(Evolutionary processes and on-line control both illustrate this latter point. An *evolving* taxon must steadily increase its fitness (average performance) if it is not to have its search ended by extinction. Similarly an on-line control system must at least maintain its current performance while searching for better, if unreasonable costs are to be avoided.)

At this point we know how an algorithm should proceed if it is to optimize the expected value of samples drawn from an arbitrary finite set of schemata (looked upon here as random variables with means and variances defined). Thus we have completed, for present purposes, steps (1) and (2) of the three step approach to the question about robust optimization algorithms. It remains to show by construction that there exist algorithms capable of this behavior. The algorithms we will consider, *genetic algorithms*, are broadly defined by the following four step iteration:

1) Determine $f(A_i)$ for each element in the data base $\mathcal{B}(t)$.

2) Produce a revised data base $\mathcal{B}'(t)$ which contains $cf(A_i)$ *copies* of each element A_i in the given data base $\mathcal{B}(t)$.
   (c is chosen to normalize f; e.g. $c = M/\Sigma_i f(A_i)$. The problem of making a "fractional copy", when $cf(A_i)$ is not a whole number, is handled by constructing an additonal copy with

probability equal to the fraction. Thus if $cf(A_i) = 2.4$, a third copy of A_i is made with probability 0.4.

3) Apply generalized genetic operators to the elements of $\mathcal{B}'(t)$ to produce a new data base $\mathcal{B}(t + 1)$.
(The genetic operators are conservative in the sense that $\mathcal{B}(t + 1)$ will have the same number of elements as $\mathcal{B}'(t)$. In general the elements of $\mathcal{B}(t + 1)$ will be different from those in $\mathcal{B}(t)$.)

4) Return to step (1).

To determine the effect of this algorithm on schemata let us first consider the effect of the first two steps. Let ξ be a schema with a set of instances $\mathcal{B}_\xi(t)$ in $\mathcal{B}(t)$. After step (2) there will be

$$c\Sigma_{A_i \varepsilon \mathcal{B}_\xi} f(A_i)$$

instances of ξ in $\mathcal{B}'(t)$. The average value of f over instances of ξ in $\mathcal{B}(t)$ is, by definition,

$$\hat{f}_\xi = \Sigma_{A_i \varepsilon \mathcal{B}_\xi} f(A_i)/|\mathcal{B}_\xi| ,$$

where $|\mathcal{B}_\xi|$ is the number of instances of ξ in $\mathcal{B}(t)$. Note that $|\mathcal{B}_\xi|/|\mathcal{B}| = |\mathcal{B}_\xi|/M = P_\xi$, the proportion of the instances of ξ in $\mathcal{B}(t)$. Similarly the proportion of instances of ξ in $\mathcal{B}'(t)$ is

$$\begin{aligned} P'(\xi) &= c\Sigma_{A_i \varepsilon \mathcal{B}_\xi} f(A_i)/c\Sigma_{A_i \varepsilon \mathcal{B}} f(A_i) \\ &= [\hat{f}_\xi \cdot |\mathcal{B}_\xi|]/[\hat{f} \cdot |\mathcal{B}|] \\ &= (\hat{f}_\xi/\hat{f})(|\mathcal{B}_\xi|/|\mathcal{B}|) = (\hat{f}_\xi/\hat{f})P_\xi \end{aligned}$$

Thus, the change in the proportion of ξ is

$$\begin{aligned} \Delta P(\xi) = P'(\xi) - P(\xi) &= [(\hat{f}_\xi/\hat{f}) - 1]P_\xi \\ &= [\hat{f}_\xi - \hat{f}]P_\xi/\hat{f} . \end{aligned}$$

This is just the action desired. The only problem is that steps (1) & (2) introduce no *new* instances of ξ. This is where step (3) enters.

Genetic operators produce new points while only producing minor perturbations in the action of the first two steps (at least for *schemata* with two or more instances in the data base). That is, the operators generate new instances of the schemata while assuring the critical rate of increase

$$\Delta P_\xi = (\hat{f}_\xi - \hat{f})P_\xi(1 - \varepsilon_\xi)/\hat{f}$$

where ε_ξ is close to zero for schemata with multiple instances.

9

The most important of the genetic operators is *crossover*. It acts on the data base to produce instances of schemata not previously present in the data base (thus providing new subsets for testing) and it provides new instances of schemata already present (thus increasing confidence in the estimates of the associated f_ξ). Moreover *each* application of this operator affects a total of $2 \cdot 2^\ell$ schemata in one of these two ways. The operator, a stochastic operator, is defined as follows:

1) Two elements $A = (a_1, a_2, \ldots, a_\ell)$ and $A' = (a'_1, a'_2, \ldots, a'_\ell)$ are selected at random (all elements equilikely) from the revised data base $B'(t)$.

2) A number x is selected with uniform probability from the interval $1 \leq x < \ell$.

3) Two new elements are formed from A and A' by exchanging the attributes of index exceeding x, yielding
$A'' = (a_1, \ldots, a_x, a'_{x+1}, \ldots, a'_\ell)$
and
$A''' = (a'_1, \ldots, a'_x, a_{x+1}, \ldots, a_\ell)$.

By considering the 2^ℓ schemata instanced by A, such as $a_1\square\ldots\square$ and $\square\ldots\square\, a_x a_{x+1}\square\ldots\square$, the reader can easily verify the effects of crossover in generating new schemata and new instances. By determining the probability that crossover will destroy an instance of a given schema, without generating a new instance of the same schema, the reader can determine the perturbation ε_ξ on step (2) of the overall algorithm. The details of these considerations, and the effects of other generalized genetic operators, can be found in Chapter 7 of reference [1].

In a data base of M ℓ-tuples randomly chosen from A ($M \ll 2^{\ell/4}$) there will be about $M^2 \cdot 2^{\ell/2}$ schemata with two or more instances. A genetic algorithm will simultaneously process the majority of these *schemata* in the way dictated by the just-discussed optimization procedure. It does this even though it manipulates elements of the data base rather than directly manipulating schemata. This amounts to a "speed-up" of about $M \cdot 2^{\ell/2}$ over any attempt to treat the schemata one-at-a-time. More specifically, let M_ξ be the number of instances of ξ in the data base, so that $P_\xi = M_\xi/M$ is the proportion of instances of ξ's in the data base. For most ξ for which $M_\xi > 1$, as indicated above, P_ξ will change according to

$$\Delta P_\xi = [\hat{f}_\xi(t) - \hat{f}(t)]P_\xi(t)(1 - \varepsilon_\xi)/\hat{f}(t)$$

under a genetic algorithm. (See Lemma 7.2 of reference [1]).

It's startling (at least it was to me) that each time a genetic algorithm processes an element of the data base, it implicitly processes nearly $M \cdot 2^{\ell/2}$ schemata in the way specified by the optimization heuristic. This *intrinsic parallelism* is a great asset in confronting high dimensional, multi-optima problems. Genetic algorithms possess an additional property of importance. After a few iterations of the algorithm, most schemata with two or more instances in the data base $\mathcal{B}(t)$, are ranked so that $\hat{f}_\xi > \hat{f}_{\xi'}$ implies that ξ has more instances in $\mathcal{B}(t)$ than ξ'. That is, each schema ξ in $\mathcal{B}(t)$ is assigned a rank between 1 and M (recalling that M is the size of $\mathcal{B}$) and this rank can be determined for ξ by simply counting the number of instances of ξ in $\mathcal{B}(t)$. (See Section 7.5 of reference [1].) Thus the genetic algorithm also compactly stores, and provides ready access to, information about a great many schemata -- M ℓ-tuples store the relative values of nearly $M \cdot 2^{\ell/2}$ schemata.

II. Applications

At this point we have completed the theoretical discussion of a general approach to optimization in highly interactive systems. It remains to interpret these results for challenging, realistic examples of such systems. Here we will briefly consider use of these tools to study two difficult problems involving highly interactive systems:

1) construction of models of cognition capable of using learning, transfer of learning, and generalization to generate realistic behavior in complex environments;

2) study of conditions adequate for the emergence of self-replicating systems from pre-biotic homogeneous media ("organic soups").

Each of these studies involves a substantial ongoing effort (documented in part by reference [2] and reference [3], respectively).

Let us begin with the models of cognition. The object is to generate response-directing classifications of the environment which have as much generality as is compatible with consistent goal-directed performance. On this view, it seems natural to equate the system's repertoire of classifiers at time t with the data base $\mathcal{B}(t)$. We will think of the system as being endowed with a set of sensors or detectors δ_i, $i = 1,\ldots,\ell$, which variously react to particular ("elemental") environmental or internal conditions. Each classifier is represented in terms of the conditions it puts on these sensors. The classifier only directs behavior when these conditions are more closely met for it than for any of its competitors in $\mathcal{B}(t)$. When this occurs the classifier is said to be *activated* and an associated set of responses is invoked. If each sensor is thought of as being either "on" (1) or "off" (0) at each moment, then the representation can be given in terms of an ℓ-tuple over the set {1,0,#}, where # at position i means that the classifier places no condition on the ith sensor. For example, 1#0##...# indicates that the classifier "looks for" the first sensor to be "on" and the third "off", the other sensor conditions being a matter of indifference to it. Each classifier thus reacts to a certain subset of the set of all possible sensor conditions.

11

The key to the operation of this system is predictive consistency. It will be the object of the optimizing algorithm to optimize this property (relative to the system's goals). The great advantage of this approach is that it allows the system to use the large flux of information in the intervals between overt goal attainment. A prediction (say of some environmental condition) either is subsequently verified or else falsified. The optimizing algorithm can use this information to improve the classifiers, even in the absense of payoff (reward for attaining a goal).

To implement this approach, we will associate with each classifier a general description of the situation ("landmark") to be expected, at some future time, if it is activated. (This description is of a subset of sensor conditions, again defined in terms of $\{1,0,\#\}$). The frequency with which this prediction is verified determines (iteratively) a performance figure, f, for the classifier. Omitting much detail, it is essentially this value f(A), for classifier $A \varepsilon \mathcal{B}(t)$, which is used by the genetic algorithm. The algorithm, thus, modifies $\mathcal{B}(t)$ to optimize predictive consistency as measured by f. (A complete recursive definition of a version of this cognitive system, including the definition of f, is given in reference [2]. It is perhaps worth emphasizing that the genetic algorithm serves as a learning procedure for "on-line" modification of the cognitive system; it is *not* being used to generate successive generations of cognitive systems.)

A final, and important, property stems from the presence, among the δ_i, of sensors for internal conditions. Activation of one classifier $A \varepsilon \mathcal{B}(t)$ can (via appropriate "internal" sensors) influence the liklihood of some other classifier $A' \varepsilon \mathcal{B}(t)$ being activated subsequently. This makes possible co-ordinated action sequences and selection of these sequences (for high f) by the genetic algorithm. It also makes possible the construction (by learning) of a cognitive map (internal model) of the environment. This map records the effects of differing responses relative to various common environmental conditions (as recorded by ℓ-tuples over $\{1,0,\#\}$). It is implemented by an appropriate *set* of classifiers. Each classifier in the set "looks for" the same environmental condition, but each generates a different response. Each also activates (via internal sensors) another set of classifiers (similarly organized) corresponding to the condition expected as a result of that response. Lookahead -- a search for the expected consequences of various lines of response -- thus becomes possible and subject to revision by the genetic algorithm.

The intrinsic parallelism of the genetic algorithm assures extensive use of all experience to shape behavior and the cognitive map. The schemata are now hyperplanes in the space $\{1,0,\#\}^{\ell}$. (E.g. 1□#□□0□...□ corresponds to the *set* of all classifiers demanding a 1 from sensor 1, being indifferent to sensor 3, and demanding a 0 from sensor 6). Various combinations ξ of 1,0,#, corresponding to these hyperplanes, are ranked in $\mathcal{B}(t)$ as dictated by the algorithm. Those combinations ξ which promote greater predictive consistency (exhibit higher $\hat{f}_{\xi}$) quickly predominate in the system's behavioral repertoire (the classifiers in $\mathcal{B}(t)$). Again, each iteration of the algorithm revises the rank of approximately $M \cdot 2^{\ell/2}$ combinations -- a great advantage to a system which must perform "on-line".

As a last point, note that the theorems concerning genetic algorithms enable us to prove theorems about the rate of learning of these systems. Such theorems (unpublished) have already been established for 2-dimensional environments where the goals are uniformly distributed with a goal-to-goal distance 2D. Initially the system performs a random walk, taking an average of D^2 units of time to reach a goal from a mid-point; learning reduces this time to D. The theorems indicate the expected time to the first increments in performance (usually from D^2 to $(D - 1)^2$) and show that subsequent increments will be achieved more rapidly. Simulations are underway to examine behavior in more complex environments.

The second of the studies, of the emergence of self-replicating systems, uses the genetic algorithm in a more obvious way. Reproduction of the basic structures in $\mathcal{A}$ (interpreted as strings of "elements" or "molecules") is a directly interpretable part of the process. Optimization begins in this case with a "search" for structures which have a reproduction rate different from zero. Emergence time -- the expected elapsed time until a self-replicating structure is generated -- plays a critical role. If the emergence time is unreasonably long, the model is valueless.

The approach is via a two step procedure. First we define a range of model "universes" having abstract counterparts of basic kinetic and biological operators (such as diffusion and enzymes). Then we determine the probabilities of emergence of self-replicating systems under homogeneous (randomized) initial conditions.

The model universes are unabashedly artificial, but straightforward elaborations should yield fair approximations, and simulations, of realistic situations, without substantially altering the basic theorems. In broad outline, the model universes have the following features.

1) The geometry is discrete and uniform, each point in the space having its immediate neighbors arrayed identically and symmetrically (typically, a rectangular grid).

2) Identical laws hold at all points in the space. (Technically, the result of (1) and (2) is a kind of cellular automaton).

3) Each point in the space can be occupied at any time by at most one "element". The elements amount to a small alphabet (cf. nucleotides or amino acids) from which the structures of interest are defined.

4) Elements are treated as persistent and mobile. That is,

elements are neither created nor destroyed in the space, but they may move (spontaneously, cf. diffusion) from point to point.

5) Elements in adjacent cells may be linked (i.e. they may form bonds), constraining them to move as a unit. The structures of interest correspond to sets of linked elements. (Technically, in the cellular automaton, the state of each cell is given by the element occupying it together with the types of links, if any, that element has with elements in adjacent cells).

6) All elements (and linked sets of elements) are subjected to the same "kinetic" operators which incessantly and randomly shift elements (and linked sets) from point to point (cf. diffusion), and which make and break linkages at random (cf. activation).

7) Some definable structures correspond to biological operators (such as enzymes) which can affect local rates of linkage formation ("bonding"). (Technically, (6) and (7) together define the transition function of the cellular automaton. It is somewhat unusual to define the transition function in terms of local operators, but this corresponds closely to the practices of physics).

8) Initial conditions are randomized homogeneous mixtures of elements and linkages. (I.e. the initial state of each cell in the cellular automaton is assigned at random).

The linked elements corresponding to the biological operators of (7) are extremely rare under the initial conditions. In effect, the universe evolves as if it were only under the influence of the "kinetic" or *primitive* operators of (6) -- an "inorganic" universe. Occasionally, the primitive operators generate a linked set of elements interpretable as a "biological" or *emergent operator*. These emergent operators in turn produce (or encourage) linked sets which would be quite rare under the action of primitive operators alone. The emergent operators are generally very context sensitive, and hence quite specific in the combinations they encourage.

Under these influences, the model universes exhibit three properties we would expect of interesting, evolving universes:

1) The stability of a structure (particularly its resistence to being broken down by the primitive operators) is a primary determinant of its density (probability of occurrence) in any given region. (For example, consider two structures A_1 and A_2 which have probabilities $p_1 = p_2$ of formation, but probabilities $q_1 = 0.1\ q_2$ of being broken down. In a space operating strictly according to these probabilities, the density of A_1 will be 10 times that of A_2).

2) The density of a structure is a primary determinant of the

frequency with which that structure is operated upon (serves as an operand). (That is, structures, including emergent operators, are being continually "shuffled" by primitive operators. Thus, of two structures satisfying a given argument, the one with high density is the more likely to serve as operand).

3) There exist operators in the universe which, when applied to a structure (operand), pass on substantial characteristics (linked subsets) of the operand to the resulting structure (pre-biotic "inheritance").

Taken together, these properties imply that any linked subset of elements which is positively correlated with above-average stability will tend toward above-average density in the space. This leads to "lines of succession" and, because elements are conserved, a transfer of elements from less stable substructures (schemata) to more stable substructures (schemata). The net effect is that different structures in A have different (implicitly defined) "reproduction" rates. As a consequence *intrinsic parallelism* plays a key role in assuring the rapid emergence of a variety of ultra-stable substructures (schemata). These in turn combine to yield autocatalytic structures which are even more stable because they actively encourage the generation of some of their own substructures. An autocatalytic structure which generates all of its own substructures is, by definition, self-replicating.

In simple versions of this universe it is easy to constructively define a self-replicating structure. In one version, 60 cells suffice to define a complete self-replicating structure. Even so, *if* we parallel the usual argument, we find that spontaneous emergence of this self-replicating structure is highly improbable. Calculations, based on a series of theorems about such cellular spaces (see reference [3]), show that the emergence time in a region of 10^4 cells is about 1.4×10^{43} time steps. If we take a time-step to be 1 millisecond, this is an emergence time of $4 \cdot 10^{30}$ centuries! This argument assumes that a self-replicating structure can only arise in one fell swoop via a random meeting of the component elements, a meeting brought about by the kinetic operators. The argument is like that which uses the mammalian eye as an argument against Darwinian evolution. It claims that the eye must emerge by a single fortuitous combination of mutations, because there are no intermediate forms with selective advantage.

In fact, as Darwin carefully shows, there *are* advantageous intermediate forms leading to the eye. Similarly, here, there are steps leading to the self-replicating structure which build upon each other and vastly reduce the emergence time.

Briefly, certain structures, which are components of the self-replicating structure, exhibit a "catalytic" action. They increase, locally, the rate of formation of other parts of the structure. These parts, because of their increased density, interact with high probability to produce more "catalytic" components. Taking these considerations into account, it is possible to show that there is a "seed" structure which, *upon emergence*, will generate the rest of the self-replicating structure in a few thousand time-steps. The emergence time for the "seed" structure is only 4.4×10^8 time-steps. (Again see reference [3] for details). With a 1 millisecond time step, this is 125 hours compared to the earlier $4 \cdot 10^{30}$ centuries!

This paper's main purpose has been to introduce the reader to the complexities, and possibilities, inherent in the study of highly interactive systems. Such systems are pervasive. They occur at critical points in psychology ("learning"), biology ("adaptation"), economics ("optimal planning"), control theory ("adaptive control"), sampling ("efficient inference"), and in the computer sciences generally ("function optimization", "artificial intelligence", etc.). It is my hope that the reader will be encouraged to view this diverse collection as a single, fundamental problem appearing in a variety of guises. Even more, I hope that he will entertain the view that these problems need not be tackled piecemeal and afresh each time encountered. A panacea (or at least a megacea) may exist.

Notes

[1]Much of the research reported here was supported by National Science Foundation Grant DCR71-01997.

Printed in the United States
By Bookmasters